CAMBRIDGE TRACTS IN MATHEMATICS

General Editors

B. BOLLOBAS, F. KIRWAN, P. SARNAK, C. T. C. WALL

132   Character Sums with Exponential Functions and their Applications

Sergei V. Konyagin

Moscow State University

Igor E. Shparlinski

Macquarie University

Character Sums with Exponential Functions and their Applications

CAMBRIDGE
UNIVERSITY PRESS

PUBLISHED BY THE PRESS SYNDICATE OF THE UNIVERSITY OF CAMBRIDGE
The Pitt Building, Trumpington Street, Cambridge, United Kingdom

CAMBRIDGE UNIVERSITY PRESS
The Edinburgh Building, Cambridge CB2 2RU, UK   www.cup.cam.ac.uk
40 West 20th Street, New York, NY 10011-4211, USA   www.cup.org
10 Stamford Road, Oakleigh, Melbourne 3166, Australia
Ruiz de Alarcón 13, 28014, Madrid, Spain

First published 1999

Printed in the United Kingdom at the University Press, Cambridge

Typeface Times 10/13pt.   *System* LATEX 2_ε [DBD]

A catalogue record of this book is available from the British Library

Library of Congress Cataloguing in Publication data
Koniagin, S. V. (Sergeĭ Vladimirovich)
Character sums with exponential functions and their applications/Sergei V. Konyagin
and Igor E. Shparlinski.
p. cm. – (Cambridge studies in advanced mathematics: 24)
"June 23, 1998."
Includes bibliographic references.
ISBN 0 521 64263 9 (hc.)
1. Exponential sums. I. Shparlinski, Igor E. II. Title.
QA246.7.K66 1999
512'.73–dc21 99-19601 CIP

ISBN 0 521 64263 9   hardback

#40776997

Contents

Preface

In this book, we consider various questions related to the distribution of integer powers g^x of some integer $g > 1$ modulo a prime number p with $\gcd(g, p) = 1$. Possible applications where such results play a central role include, but are not limited to, linear congruential pseudo-random number generators, algebraic number theory, the theory of function fields over a finite field, complexity theory, cryptography, and coding theory.

We also consider similar questions in a more general setting related to the distribution of elements of a finitely generated multiplicative group V with r generators in an algebraic number field $\mathbb{K}$ of degree n over $\mathbb{Q}$ modulo an integer ideal $\mathfrak{q}$.

Acknowledgement

The authors are grateful to Alf van der Poorten for his help, support and valuable advice.

The authors also wish to thank Hugh Montgomery and Martin Tompa for a helpful discussion of the problems considered in Chapter 15 and Francesco Pappalardi for information about his result [72] and many fruitful discussions of some problems. The authors are thankful to Roger Heath-Brown for providing us with some of his unpublished results. Corey Powell kindly provided a preliminary version of [73].

The authors would like to thank Gang Yu who helped them to find some mistakes in a preliminary version of Chapter 6.

The research of the first author has been supported by the Grant DMS 9304580 from the NSF.

The research of the second author has been supported in part by the Grant A69700294 of the Australian Research Council.

Part one
Preliminaries

1

Introduction

The main subject of this book can be described as a study of various properties of the distribution of integer powers g^x of some integer $g > 1$ modulo a prime number p with $\gcd(g, p) = 1$. We are also interested in applications of such results to various problems. In particular, we consider several well-known problems from algebraic number theory, the theory of function fields over a finite field, complexity theory, the theory of linear congruential pseudo-random number generators, cryptography, and coding theory.

To describe more precisely the type of questions which we study in this book and which arise in the aforementioned applications, let us denote by t the multiplicative order modulo p of an integer $g > 1$ with $\gcd(g, p) = 1$.

For $(a, p) = 1$, $1 \le N \le t$, $0 \le M < p$, $1 \le H \le p$, we denote by $T_a(N, M, H)$ the number of solutions of

$$ag^x \equiv M + u \pmod{p}, \qquad 1 \le x \le N, \ 1 \le u \le H.$$

Typically, the aforementioned problems lead to one of the following questions about the distribution of residues of an exponential function.

- What is the largest value of $|T_a(N, M, H) - NH/p|$ over all $a = 1, \ldots,$ $p - 1$ and $M = 0, \ldots, p - 1$?
- What are the restrictions on N and H, under which $T_a(N, M, H) > 0$ for every M?
- For how many integers i, $0 \le i \le p/H - 1$, is $T_a(N, iH, H) > 0$?
- What is the largest value of H (as a function of N) for which $T_a(N, M, H) = 0$ for some M?
- What is the smallest value of H (as a function of N) for which $T_a(N, M, H) = N$ for some M?

These questions may be asked:

- for all $a \in \mathbb{F}_p^*$;
- for some special a, say $a = 1$;
- for 'almost all' $a \in \mathbb{F}_p^*$;
- for at least one 'good' $a \in \mathbb{F}_p^*$;

and in several other cases.

Similar questions can be considered modulo composite numbers and, even more generally, for finitely generated multiplicative groups of algebraic number fields which are reduced modulo an integer ideal of that field.

Looking at the subjects that interest us, it should not be a big surprise that our main tool is various bounds for character sums. Thus we start this book with a collection of known relevant bounds as well as several new ones. In particular, we obtain new bounds of Gaussian sums. Indeed, it is easy to see that many questions about the distribution of g^x modulo p are equivalent to similar questions about the distribution of the x^n modulo p, where $n = (p - 1)/t$, and this leads to Gaussian sums. Certainly the last subject is of great independent interest and we consider this topic as well. Then we present a series of new results on the structure of multiplicative shifts of multiplicative subgroups and arbitrary subsets of $\mathbb{F}_p^*$. In subsequent chapters, we give a wide spectrum of applications of these basic results.

As we have mentioned, studying the distribution of residues g^x modulo p is our central interest and is most important for the majority of our applications. Nevertheless, in some cases we need to consider the more general situation with finitely generated groups in algebraic number fields. This is why we formulate our main results concerning bounds of exponential sums in terms of such groups (even if the actual result is applicable only to the $g^x \pmod{p}$). The reader who is not interested in applications to algebraic number fields may always assume that 'integer ideal' means 'integer', 'prime ideal' means 'prime number', 'algebraic number field K' means just 'field of rationals $\mathbb{Q}$', finitely generated groups have rational integer generators, and so on.

There are also some technical reasons to work in a more general setting for arbitrary algebraic number fields. In fact, some of our results are proved (and formulated, of course) for the basic case of $g^x \pmod{p}$. Nevertheless, we believe they hold in the full generality. Obtaining such generalizations would be very important for a number of applications. In particular, we believe that in many of our statements, the words 'let $\mathfrak{p}$ be a prime ideal of first degree' (which essentially refer to the distribution modulo p) can be simplified to just 'let $\mathfrak{p}$ be a prime ideal'. We should remark that, as far as we can see, such generalizations will not be simple exercises but will require some new ideas.

In fact we hope that such new ideas could turn out to be useful for obtaining further results about the distribution of g^x modulo p as well.

Let $\mathbb{K}$ be an algebraic number field of degree n over the field of rational numbers $\mathbb{Q}$, and let $\mathbb{Z}_\mathbb{K}$ be its ring of integers. For an integer ideal q, we denote by Λ_q the residue ring modulo q and by Λ_q^* the multiplicative group of units of this ring.

Given a finitely generated multiplicative group V of $\mathbb{K}$

$$V = \{\lambda_1^{x_1} \ldots \lambda_r^{x_r} \; : \; x_1, \ldots, x_r \in \mathbb{Z}\},$$

we denote its reduction modulo q by V_q. We shall always suppose that the generators $\lambda_1, \ldots, \lambda_r$ are multiplicatively independent.

There are a great many results on the behavior of groups V in $\mathbb{K}$ [29, 30, 13]. Here we concentrate on their reductions V_q. In the simplest, but probably the most important case, when $\mathbb{K} = \mathbb{Q}$ and $r = 1$, this is a classical question about the distribution of residues of an exponential function equivalent to considering the quality of the linear congruential pseudo-random number generator [37, 67, 69]. We shall consider this and other applications which rely on results which are not so widely known concerning the distribution of V_q in Λ_q.

As we have mentioned, in many situations it is enough to study the case $\mathbb{K} = \mathbb{Q}, r = 1$ and moreover q = p is a rational prime number.

Such applications include but are not limited to:

- Egami's question about smallest norm representatives of the residue classes modulo q and Euclid's algorithm [12, 79];
- Prediction of the $1/M$-pseudo-random number generator of Blum, Blum, and Shub [6] and the linear congruential generator [16];
- Girstmair's problem about the relative class number of subfields of cyclotomic fields [20, 21, 22] and Myerson's problem about Gaussian periods [62, 63];
- Kodama's question about supersingular hyperelliptic curves [64, 68, 92, 93];
- Tompa's question about lower bounds for the QuickSort algorithm using a linear congruential pseudo-random number generator [36, 90];
- Lenstra's constants modulo q and Győry's arithmetical graphs [29, 30, 48, 49, 65, 70];
- Estimating the dimension of BCH codes [5, 54];
- Robinson's question about small mth roots modulo p [75];
- Håstad, Lagarias, and Odlyzko's question about the average value of smallest elements in multiplicative translations of sets modulo p [31];
- Niederreiter's problem about the multiplier of linear congruential pseudo-random number generators [67, 69];

- Stechkin's question about the constant in the estimate of Gaussian sums with arbitrary denominators [86];
- Odlyzko and Stanley's problem about 0, 1-solutions of a certain congruence modulo p [71].

It is easy to extend the list of problems which are related to questions on the distribution or residues of finitely generated groups. As an example, we note papers [9, 23] where links to the weight distribution of arithmetic codes are displayed.

Another example is paper [51] where some properties of finitely generated groups were used to study certain algebraic questions. All these properties (combinations of Artin's conjecture and Tchebotarev's density theorem) were established (under the Extended Riemann Hypothesis, of course) in [60] which was motivated by [51].

Many other problems about the minimal polynomials of Gaussian periods (over rationals as well as over finite fields) are considered in [19, 24, 25, 26, 27, 28]. We also refer to [88, 89] for good expositions of various basic properties of Gaussian periods and related questions. Perhaps the methods of the present book can be applied to some of them. Indeed, in Chapter 10 we consider the problem about the norm of Gaussian periods. A more general question of computing their minimal polynomials is of great interest too (for details see the papers above). It turns out that several higher coefficients can be expressed in terms of the numbers $R(k, t, p)$, $k = 1, 2, \ldots$, of solutions of the equations

$$g_1 + \cdots + g_k = 0, \qquad g_1, \ldots, g_k \in G_t,$$

(we follow the notation of Chapter 10). Thus using various bounds of exponential sums, one can estimate (or even find an asymptotic formula for) $T(k, t, p)$ and then apply them to studying higher coefficients.

Of course any improvement of bounds of exponential sums used in this book would entail further progress in this area. The same is true for any improvement of Lemma 9.7.

Also, many questions about the distribution of residues of multiplicative groups can be reformulated in the dual form as questions about the distribution of indices and therefore bounds of multiplicative character sums, including the celebrated Burgess estimate, can be used. For example, see the remarks made in Chapter 15 and Chapter 7, and another example of using character sums in this kind of question is given in [23].

Generally, we do not try to extract all possible results accessible by our methods, nor do we try to get the best possible values of some (important) constants. Rather, we attempt to demonstrate different approaches from various areas of mathematics in one attack on certain problems. One of the examples is Theorem 6.7 which is based on some delicate combinations of tools from mathematical analysis, geometry of numbers, and algebraic geometry. We pose several problems of different levels of difficulty. Some of them can probably be solved within the framework of this book, others will require some radically new ideas (although in general we try to avoid posing hopeless problems). We would like to believe that this book will stimulate further research in this very important and mathematically attractive area.

Finally, we stress that it would be interesting to consider similar questions for some other groups, say for finitely generated matrix groups, for groups of points on elliptic curves, or for finitely generated groups in function fields.

2

Notation and Auxiliary Results

Here we collect some notation and useful facts which we use repeatedly throughout this book.

We denote by $\log x$ the binary logarithm of x and by $\ln x$ the natural logarithm of x.

Several of our estimates include iterations of logarithmic functions and do not make any sense for some values of arguments. To save space and avoid using expressions like $\log \max\{2, \log \max\{2, k\}\}$, we define

$$\text{Log } x = \max\{2, \log x\}, \qquad \text{Ln } x = \max\{2, \ln x\}.$$

For a complex $z \in \mathbb{C}$, we denote by $\Re z$ its real part.

For a prime number p and an integer $a \neq 0$, we denote by $\text{ord}_p \, a$ the p-adic order of a, that is the largest power of p which divides a.

For brevity, we set

$$\mathbf{e}(z) = \exp(2\pi i z).$$

As usual, $\pi(x)$ denotes the number of prime numbers which do not exceed x and $\pi(x, k, l)$ denotes the number of primes which do not exceed x and are congruent to l modulo k.

We also make use of the following estimates:

$$k - 1 \geq \varphi(k) \gg \frac{k}{\text{Ln ln } k}, \qquad \omega(k) \ll \frac{\ln k}{\text{Ln ln } k},$$

where $\varphi(k)$ is the Euler function and $\omega(k)$ is the number of prime divisors of integer $k \geq 2$, and

$$\tau(k) \leq \exp\left((\ln 2 + o(1)) \frac{\ln k}{\ln \ln k}\right), \qquad k \to \infty,$$

where $\tau(k)$ is the number of integer positive divisors of $k \geq 2$.

8

They easily follow from the Prime Number Theorem and can be found in [74] and many other sources.

For an element ϑ of a ring $\mathcal{R}$ we define the multiplicative order of ϑ as the smallest integer $t > 0$ for which $\vartheta^t = 1$, if such an integer exists, otherwise the multiplicative order is undefined. It is easy to see that if $\mathcal{R}$ is a finite ring and ϑ is not a zero divisor that the multiplicative order is always defined.

For an algebraic extension $\mathbb{L}$ of a field $\mathbb{K}$, $\mathrm{Tr}_{\mathbb{L}/\mathbb{K}}(\alpha)$ and $\mathrm{Nm}_{\mathbb{L}/\mathbb{K}}(\alpha)$ denote the trace and the norm of $\alpha \in \mathbb{L}$ in $\mathbb{K}$, respectively. That is,

$$\mathrm{Tr}_{\mathbb{L}/\mathbb{K}}(\alpha) = \sum_{i=1}^{s} \sigma_i(\alpha) \quad \text{and} \quad \mathrm{Nm}_{\mathbb{L}/\mathbb{K}}(\alpha) = \prod_{i=1}^{s} \sigma_i(\alpha),$$

where σ_i, $i = 1, \ldots, s$, are distinct embeddings of $\mathbb{L}$ into the algebraic closure of $\mathbb{K}$, $s = [\mathbb{L} : \mathbb{K}]$. It is easy to verify that for a chain of extensions $\mathbb{F} \subseteq \mathbb{K} \subseteq \mathbb{L}$ we have

$$\mathrm{Tr}_{\mathbb{L}/\mathbb{F}}(\alpha) = \mathrm{Tr}_{\mathbb{L}/\mathbb{K}}\left(\mathrm{Tr}_{\mathbb{K}/\mathbb{F}}(\alpha)\right)$$

and

$$\mathrm{Nm}_{\mathbb{L}/\mathbb{F}}(\alpha) = \mathrm{Nm}_{\mathbb{L}/\mathbb{K}}\left(\mathrm{Nm}_{\mathbb{K}/\mathbb{F}}(\alpha)\right).$$

Let $\mathbb{K}$ be an algebraic number field of degree n over the field of rational numbers $\mathbb{Q}$. We denote by $\mathbb{Z}_{\mathbb{K}}$ the ring of integers of $\mathbb{K}$, that is the ring of elements of $\mathbb{K}$ whose minimal polynomial over $\mathbb{Q}$ is monic.

For an integer ideal $\mathfrak{q}$, we denote by $\Lambda_{\mathfrak{q}}$ the residue ring modulo $\mathfrak{q}$ and by $\Lambda_{\mathfrak{q}}^*$ the multiplicative group of units of this ring. It is well known that $|\Lambda_{\mathfrak{q}}| = \mathrm{Nm}(\mathfrak{q})$ and actually this can be taken as a definition of $\mathrm{Nm}(\mathfrak{q})$.

For any prime ideal $\mathfrak{p}$, $\mathrm{Nm}(\mathfrak{p}) = p^d$ for some prime p and integer d, $1 \leq d \leq n$, which is called the degree of $\mathfrak{p}$.

If $\mathfrak{p}$ is a prime ideal of degree d then $\Lambda_{\mathfrak{p}} \simeq \mathbb{F}_{p^d}$, a finite field of $p^d = \mathrm{Nm}(\mathfrak{p})$ elements.

It is easy to see that $\mathfrak{p}|p$ in $\mathbb{Z}_{\mathbb{K}}$. The ideal $\mathfrak{p}$ is called ramified if $\mathfrak{p}^2|p$ and unramified otherwise.

If $\mathfrak{p}$ is an unramified prime ideal of first degree then $\Lambda_{\mathfrak{p}^k} \simeq \mathbb{Z}/(p^k)$ where $p = \mathrm{Nm}(\mathfrak{p})$.

We also have $|\Lambda_{\mathfrak{q}}^*| = \varphi(\mathfrak{q})$, where $\varphi(\mathfrak{q})$ is the Euler function in $\mathbb{Z}_{\mathbb{K}}$, which has properties very similar to these of the Euler function in $\mathbb{Z}$. For example, it is multiplicative and

$$\varphi(\mathfrak{p}^r) = \mathrm{Nm}(\mathfrak{p})^{r-1}\left(\mathrm{Nm}(\mathfrak{p}) - 1\right)$$

for any prime ideal power $\mathfrak{p}^r$.

The residue ring Λ_q has $\mathrm{Nm}(q)$ additive characters χ which are functions $\chi : \Lambda_q \to \mathbb{C}$ such that

$$\chi(z_1 + z_2) = \chi(z_1)\chi(z_2) \quad \text{and} \quad |\chi(z)| = 1$$

for any $z_1, z_2, z \in \Lambda_q$. The character χ_0 with $\chi_0(z) = 1$, $z = \Lambda_q$ is called trivial. Multiplicative characters are defined in a similar way with respect to the group Λ_q^*.

For a rational integer q the corresponding characters are of the form $\chi_a(z) = \mathbf{e}(az/q)$ for $a = 0, \ldots, q - 1$. For a prime ideal $\mathfrak{p}$ of norm $\mathrm{Nm}(\mathfrak{p}) = p^d$ the characters of $\Lambda_{\mathfrak{p}}$ are of the form

$$\chi_a(z) = \mathbf{e}\left(\mathrm{Tr}_{\mathbb{K}/\mathbb{Q}}(az)/p\right), \qquad a \in \Lambda_{\mathfrak{p}}.$$

In both cases $a = 0$ corresponds to the trivial character.

Finally we mention two our very frequently used tools. The first one is the Cauchy inequality

$$\sum_{i=1}^{N} A_i B_i \leq \left(\sum_{i=1}^{N} A_i^{\alpha}\right)^{1/\alpha} \left(\sum_{i=1}^{N} B_i^{\beta}\right)^{1/\beta}$$

which holds for any two sequences of positive numbers $A_i, B_i, i = 1, \ldots, N$ and any positive α, β with $\alpha^{-1} + \beta^{-1} = 1$.

The second one is the Hadamard inequality

$$|\det A|^2 \leq \prod_{i=1}^{N} \sum_{j=1}^{N} |a_{ij}|^2$$

for the determinant of a matrix $A = \left(a_{ij}\right)_{i,j=1}^{N}$ with complex elements.

Part two

Bounds of Character Sums

3

Bounds of Long Character Sums

Here we present estimates of character sums with sufficiently many terms.

Given a primitive additive character χ of the ring $\Lambda_{\mathfrak{q}}$, we define the sum of kth powers and the largest value of the following character sums:

$$\sigma_k(\mathfrak{q}, V) = \sum_{\alpha \in \Lambda_{\mathfrak{q}}} \left| \sum_{v \in V_{\mathfrak{q}}} \chi(\alpha v) \right|^k, \qquad S(\mathfrak{q}, V) = \max_{\alpha \in \Lambda_{\mathfrak{q}}^*} \left| \sum_{v \in V_{\mathfrak{q}}} \chi(\alpha v) \right|,$$

noting that this definition does not, of course, depend on the choice of the character χ.

For a group V and an integer ideal $\mathfrak{q}$, let us denote by $T_k(\mathfrak{q}, V)$ the number of solutions of the congruence

$$v_1 + \cdots + v_k \equiv u_1 + \cdots + u_k \pmod{\mathfrak{q}}, \qquad v_1, u_1, \ldots, v_k, u_k \in V_{\mathfrak{q}}.$$

Obviously, for any integer ideal $\mathfrak{q}$, we have

$$T_1(\mathfrak{q}, V) = |V_{\mathfrak{q}}|. \tag{3.1}$$

We also denote by $N_k(a, \mathfrak{q}, V)$ the number of solutions of the congruence

$$v_1 + \cdots + v_k \equiv a \pmod{\mathfrak{q}}, \qquad v_1, \ldots, v_k \in V_{\mathfrak{q}}.$$

These quantities play the crucial role in our estimates of character sums. In particular, we repeatedly use the following evident identities:

$$\sum_{a \in \Lambda_{\mathfrak{q}}} N_k(a, \mathfrak{q}, V) = |V_{\mathfrak{q}}|^k, \qquad \sum_{a \in \Lambda_{\mathfrak{q}}} N_k(a, \mathfrak{q}, V)^2 = T_k(\mathfrak{q}, V). \tag{3.2}$$

Very often we use the following basic feature of primitive characters. For any $\beta \in \Lambda_{\mathfrak{q}}$,

$$\sum_{\alpha \in \Lambda_{\mathfrak{q}}} \chi(\alpha \beta) = \begin{cases} 0, & \text{if } \beta \neq 0; \\ \mathrm{Nm}(\mathfrak{q}), & \text{if } \beta = 0. \end{cases} \tag{3.3}$$

13

In particular, for any set M, we have

$$\sum_{\alpha \in \Lambda_q} \left| \sum_{\mu \in M} \chi(\alpha\mu) \right|^{2k} = \mathrm{Nm}(q) \, Q_{2k}(M), \tag{3.4}$$

where $Q_{2k}(M)$ is the number of solutions of the congruence

$$\mu_1 + \cdots + \mu_k \equiv \mu_{k+1} + \cdots + \mu_{2k} \pmod{q}, \qquad \mu_1, \ldots, \mu_{2k} \in M.$$

The results of this chapter will be mainly used as bounds on exponential sums with exponential functions. Because of this and because it can be useful for readers who are not familiar with algebraic number fields and are interested only in our results for the field of rational numbers we present the above definitions and identities for that particular case. Later we will formulate our main results in this form as well.

Let p be prime and let g be an integer with $\gcd(g, p) = 1$. Denote by t the multiplicative order of g modulo p, and let $V = \{1, g, \ldots, g^{t-1}\}$. Then the definitions of $\sigma_k(q, V)$ and $S(q, V)$ can be rewritten as

$$\sigma_k(p, V) = \sum_{a \in \mathbb{F}_p} \left| \sum_{v \in V} \mathbf{e}(av/p) \right|^k, \qquad S(p, V) = \max_{a \in \mathbb{F}_p^*} \left| \sum_{v \in V} \mathbf{e}(av/p) \right|.$$

$T_k(p, V)$ is the number of solutions of the congruence

$$v_1 + \cdots + v_k \equiv u_1 + \cdots + u_k \pmod{p}, \qquad v_1, u_1, \ldots, v_k, u_k \in V,$$

and $N_k(a, p, V)$ is the number of solutions of the congruence

$$v_1 + \cdots + v_k \equiv a \pmod{p}, \qquad v_1, \ldots, v_k \in V.$$

Then the identities (3.1), (3.2), (3.3) and (3.4) take the form:

$$T_1(p, V) = |V|,$$

$$\sum_{a \in \mathbb{F}_p} N_k(a, p, V) = |V|^k, \qquad \sum_{a \in \mathbb{F}_p} N_k(a, p, V)^2 = T_k(p, V),$$

$$\sum_{a \in \mathbb{F}_p} \mathbf{e}(ab/p) = \begin{cases} 0, & \text{if } b \not\equiv 0 \pmod{p}; \\ p, & \text{if } b \equiv 0 \pmod{p}; \end{cases}$$

and

$$\sum_{a \in \mathbb{F}_p} \left| \sum_{m \in M} \mathbf{e}(am/p) \right|^{2k} = p \, Q_{2k}(M),$$

where $M \subset \mathbb{F}_p$ and $Q_{2k}(M)$ is the number of solutions of the congruence

$$m_1 + \cdots + m_k \equiv m_{k+1} + \cdots + m_{2k} \pmod{p}, \qquad m_1, \ldots, m_{2k} \in M.$$

Besides those estimates above, we repeatedly use the following elementary inequalities which are essentially Exercises 11b and 11c in Chapter 3 of [91].
The first one is

$$\left| \sum_{h_1 \le y < h_2} \mathbf{e}(ay/m) \right| \ll \min\{h_2 - h_1, m/|a|\} \tag{3.5}$$

and it holds for any integers h_1, h_2 and $0 \le |a| < m/2$. From this inequality, one easily derives that

$$\sum_{a=1}^{m-1} \left| \sum_{h_1 \le y < h_2} \mathbf{e}(ay/m) \right| = \sum_{-m/2 \le a \le (m-1)/2} \left| \sum_{h_1 \le y < h_2} \mathbf{e}(ay/m) \right| \ll m \ln m \tag{3.6}$$

for any integers h_1, h_2 with $m \ge h_2 - h_1 \ge 0$. Actually h_1, h_2 can even be any functions of a.

We repeatedly use the identities (3.3), (3.4) and the inequalities (3.5), (3.6) throughout this book (the identity (3.4) usually, but not always, with $k = 1$).

In particular, from the identity (3.4), one can see that

$$\sigma_{2k}(\mathfrak{q}, V) = \mathrm{Nm}(\mathfrak{q}) T_k(\mathfrak{q}, V).$$

We use this equation (and similar ones) in many places. For instance, in the proof of the following statement.

Lemma 3.1. *For any integer $k \ge 1$, the inequalities*

(i) $S(\mathfrak{q}, V) \le \left(\mathrm{Nm}(\mathfrak{q}) T_k(\mathfrak{q}, V)/|V_{\mathfrak{q}}| \right)^{1/2k}$,

(ii) $S(\mathfrak{q}, V) \le \mathrm{Nm}(\mathfrak{q})^{1/2k^2} T_k(\mathfrak{q}, V)^{1/k^2} |V_{\mathfrak{q}}|^{1-2/k}$,

hold.

Proof Taking into account that $N_k(a, \mathfrak{q}, V) = N_k(av, \mathfrak{q}, V)$ for any $v \in V_{\mathfrak{q}}$, we obtain

$$\left[\sum_{v \in V_{\mathfrak{q}}} \chi(\alpha v) \right]^k = \sum_{a \in \Lambda_{\mathfrak{q}}} N_k(a, \mathfrak{q}, V) \chi(\alpha a)$$

$$= \frac{1}{|V_{\mathfrak{q}}|} \sum_{a \in \Lambda_{\mathfrak{q}}} N_k(a, \mathfrak{q}, V) \sum_{v \in V_{\mathfrak{q}}} \chi(\alpha a v).$$

Let $l \geq 1$ be an integer. Now, from the identities (3.2), we derive

$$\left| \sum_{v \in V_{\mathfrak{q}}} \chi(\alpha v) \right|^{2kl}$$

$$\leq \frac{1}{|V_{\mathfrak{q}}|^{2l}} \left[\sum_{a \in \Lambda_{\mathfrak{q}}} N_k(a, \mathfrak{q}, V)^{2l/(2l-1)} \right]^{2l-1} \sum_{a \in \Lambda_{\mathfrak{q}}} \left| \sum_{v \in V_{\mathfrak{q}}} \chi(\alpha a v) \right|^{2l}$$

$$= \frac{\mathrm{Nm}(\mathfrak{q}) T_l(\mathfrak{q}, V)}{|V_{\mathfrak{q}}|^{2l}} \left[\sum_{a \in \Lambda_{\mathfrak{q}}} N_k(a, \mathfrak{q}, V)^{2l/(2l-1)} \right]^{2l-1}$$

$$\leq \frac{\mathrm{Nm}(\mathfrak{q}) T_l(\mathfrak{q}, V)}{|V_{\mathfrak{q}}|^{2l}} \left[\sum_{a \in \Lambda_{\mathfrak{q}}} N_k(a, \mathfrak{q}, V) \right]^{2l-2} \sum_{a \in \Lambda_{\mathfrak{q}}} N_k(a, \mathfrak{q}, V)^2$$

$$= \mathrm{Nm}(\mathfrak{q}) T_k(\mathfrak{q}, V) T_l(\mathfrak{q}, V) |V_{\mathfrak{q}}|^{2kl - 2k - 2l}.$$

Therefore

$$S(\mathfrak{q}, V) \leq \mathrm{Nm}(\mathfrak{q})^{1/2kl} T_k(\mathfrak{q}, V)^{1/2kl} T_l(\mathfrak{q}, V)^{1/2kl} |V_{\mathfrak{q}}|^{1 - 1/k - 1/l}. \qquad (3.7)$$

Selecting $l = 1$ and using (3.1), we obtain the first bound. Selecting $l = k$, we obtain the second bound. $\qquad \qquad \square$

Although we have proved a more general bound (3.7), for all applications, only the two partial cases, given in Lemma 3.1, are important.

The following lemma is a generalization of the bound of [17] on the number of rational points on the Fermat curve $x^n + y^n = a$ over a prime finite field which improves the Weil bound. It is shown in [33] that the same estimate (up to the value of the constant) can be obtained by the elementary method (which actually works for any field of characteristic p). In our notation, this result can be formulated as follows.

For any prime ideal $\mathfrak{p}$ of first degree with $\mathrm{Nm}(\mathfrak{p}) = p$ and such that

$$|V_{\mathfrak{p}}| < \frac{\mathrm{Nm}(\mathfrak{p}) - 1}{(\mathrm{Nm}(\mathfrak{p}) - 1)^{1/4} + 1},$$

the bound

$$N_2(a, \mathfrak{p}, V) \leq 4 |V_{\mathfrak{p}}|^{2/3} \qquad (3.8)$$

holds.

In fact here we present a more general statement, obtained by the method of [33], which we then apply to prove stronger bounds of exponential sums than those that can be derived from (3.8) alone.

The bound (3.8) follows from that statement (up to the value of the constant) provided that $|V_{\mathfrak{p}}|$ satisfies a slightly stronger restriction, namely, $|V_{\mathfrak{p}}| \leq C\, \mathrm{Nm}(\mathfrak{p})^{3/4}$, where C is any positive constant with $C < 2^{-1/4}$; in particular, one can select $C = 0.8$.

Lemma 3.2. *For any prime ideal $\mathfrak{p}$ of first degree with $\mathrm{Nm}(\mathfrak{p}) = p$, $|V_{\mathfrak{p}}| = t$, and for any positive integers B and s such that*

$$B < t, \qquad tB \leq p, \qquad s < B^3/(2t), \qquad (3.9)$$

and any elements $a_1, \ldots, a_s$ selected from pairwise distinct co-sets of the subgroup $V_{\mathfrak{p}}$ in $\Lambda_{\mathfrak{p}}^*$, that is such that $a_i/a_j \notin V_{\mathfrak{p}}$ for $1 \leq i < j \leq s$, the bound

$$\sum_{j=1}^{s} N_2(a_j, \mathfrak{p}, V) \leq \frac{2tB}{\lfloor t/B \rfloor}$$

holds.

Proof Let $A = D = \lfloor t/B \rfloor$. We shall begin by taking a polynomial $\Phi(X, Y, Z) \in \mathbb{F}_p[X, Y, Z]$, for which

$$\deg_X \Phi < A, \qquad \deg_Y \Phi < B, \qquad \deg_Z \Phi < B.$$

For $j = 1, \ldots, s$ we define the sets

$$R_j = \{x : x \in V_{\mathfrak{p}}, \, a_j - x \in V_{\mathfrak{p}}\}$$
$$S_j = \{x : a_j x \in V_{\mathfrak{p}}, \, a_j - x a_j \in V_{\mathfrak{p}}\}.$$

We see that $R_j = a_j S_j$, hence $|R_j| = |S_j|$, $j = 1, \ldots, s$. By the condition on a_j, the sets S_j are pairwise disjointed, $j = 1, \ldots, s$. Therefore, if we put

$$S = \bigcup_{j=1}^{s} S_j,$$

we obtain

$$\sum_{j=1}^{s} N_2(a_j, \mathfrak{p}, V) = \sum_{j=1}^{s} |R_j| = \sum_{j=1}^{s} |S_j| = |S|.$$

The underlying idea is then to arrange that the polynomial

$$\Psi(X) = \Phi(X, X^t, (1 - X)^t)$$

has a zero of order at least D, say, at each point $x \in S$. We will therefore be able to conclude that

$$D \sum_{j=1}^{s} N_2(a_j, \mathfrak{p}, V) \le \deg \Psi(X),$$

providing that Ψ does not vanish identically. We note that

$$\deg \Psi \le \deg_X \Phi + t \deg_Y \Phi + t \deg_Z \Phi \le A - 1 + 2t(B - 1),$$

whence

$$\sum_{j=1}^{s} N_2(a_j, \mathfrak{p}, V) \le (A - 1 + 2t(B - 1)) / D < \frac{2tB}{\lfloor t/B \rfloor},$$

providing that Ψ does not vanish.

In order for Ψ to have a zero of multiplicity at least D at a point x, we need

$$\left(\frac{d}{dX}\right)^n \Psi(X) \bigg|_{X=x} = 0 \quad \text{for} \quad n < D.$$

Since $x \ne 0, 1$ for $x \in S$, this will be equivalent to

$$\{X(1 - X)\}^n \left(\frac{d}{dX}\right)^n \Psi(X) \bigg|_{X=x} = 0. \tag{3.10}$$

We now observe that

$$X^m \left(\frac{d}{dX}\right)^m X^u = \frac{u!}{(u - m)!} X^u,$$

$$X^m \left(\frac{d}{dX}\right)^m X^{tv} = \frac{(tv)!}{(tv - m)!} X^{tv},$$

$$(1 - X)^m \left(\frac{d}{dX}\right)^m (1 - X)^{tw} = (-1)^m \frac{(tw)!}{(tw - m)!} (1 - X)^{tw}.$$

It follows that

$$\{X(1 - X)\}^k \left(\frac{d}{dX}\right)^k X^u X^{tv} (1 - X)^{tw} = P_{k,u,v,w}(X) X^{tv} (1 - X)^{tw}$$

where $P_{k,u,v,w}(X)$ either vanishes or is a polynomial of degree at most $k + u$. We therefore deduce that for any $j = 1, \dots, s$ and any $x \in S_j$, we have

$$\{X(1 - X)\}^k \left(\frac{d}{dX}\right)^k X^u X^{tv}(1 - X)^{tw}\Bigg|_{X=x} = a_j^{-t(v+w)} P_{k,u,v,w}(x)$$

because $x^t = (1 - x)^t = a_j^{-t}$ for $x \in S_j$.

We now write

$$\Phi(X, Y, Z) = \sum_{u,v,w} \lambda_{u,v,w} X^u Y^v Z^w$$

and

$$P_{k,j}(X) = \sum_{u,v,w} \lambda_{u,v,w} a_j^{-t(v+w)} P_{k,u,v,w}(X),$$

so that $\deg P_{k,j}(X) < A + k$ and

$$\{X(1 - X)\}^k \left(\frac{d}{dX}\right)^k \Phi(X, X^t, (1 - X)^t)\Bigg|_{X=x} = P_{k,j}(x)$$

for any $x \in S_j$. We shall arrange, by appropriate choice of the coefficients $\lambda_{u,v,w}$, that $P_{k,j}(X)$ vanishes identically for $k < D$. This will ensure that (3.10) holds at every point $x \in S$. Each polynomial $P_{k,j}(X)$ has at most $A + k \leq A + D$ coefficients, which are linear forms in the original $\lambda_{u,v,w}$. Thus if

$$sD(A + D) < AB^2, \qquad (3.11)$$

there will be a set of coefficients $\lambda_{u,v,w}$, not all zero, for which the polynomials $P_{k,j}(X)$ vanish for all $k < D$. But, by (3.9),

$$sD(A + D) = 2sA^2 \leq 2sAt/B < AB^2,$$

and (3.11) holds.

We must now consider whether the polynomial $\Phi(X, X^t, (1 - X)^t)$ can vanish if $\Phi(X, Y, Z)$ does not. We shall write

$$\Phi(X, Y, Z) = \sum_w \Phi_w(X, Y)Z^w,$$

and take w_0 to be the smallest value of w for which $\Phi_w(X, Y)$ is not identically zero. It follows that

$$\Phi(X, X^t, (1 - X)^t) = (1 - X)^{tw_0} \sum_{w_0 \leq w < B} \Phi_w(X, X^t)(1 - X)^{t(w - w_0)},$$

so that if $\Phi(X, X^t, (1 - X)^t)$ is identically zero, we must have

$$\Phi_{w_0}(X, X^t) \equiv 0 \pmod{(1 - X)^t}. \tag{3.12}$$

It is easy to see that if a polynomial $f(X) \in \mathbb{F}_p[X]$ of degree $\deg f < p$ is a sum of $N \geq 1$ distinct monomials, then $(1 - X)^N$ cannot divide $f(X)$. Indeed, it can be shown by induction on N. The case $N = 1$ is trivial. Now suppose that $N > 1$ and let

$$f(X) = \sum_w c_w X^w$$

where w runs over N distinct values. Then the polynomial

$$g(X) = Xf'(X) - Wf(X) = \sum_w c_w (w - W) X^w,$$

where $W = \deg f$, contains exactly $N - 1$ terms. We then see that if $(1 - X)^N$ divides $f(X)$, then $(1 - X)^{N-1}$ divides $g(X)$ contrary to our induction hypothesis.

We have

$$\deg \Phi_{w_0}(X, X^t) \leq A - 1 + t(B - 1) < tB.$$

Therefore the congruence (3.12) is impossible, providing that

$$AB \leq t \qquad \text{and} \qquad tB \leq p.$$

These inequalities hold by (3.9), which gives the desired result. $\qquad\square$

It is not difficult to check that for sufficiently large t with $t \leq 0.8 p^{3/4}$, $B = \lfloor (2t)^{1/3} \rfloor + 1$, and $s = 1$, the conditions (3.9) hold and we get the estimate (3.8) up to the value of the constant.

Now we can prove the following version of the corresponding statement from [33].

Lemma 3.3. *For any prime ideal $\mathfrak{p}$ of first degree with $\mathrm{Nm}(\mathfrak{p}) = p$ and such that*

$$|V_\mathfrak{p}| < 0.7(\mathrm{Nm}(\mathfrak{p}))^{2/3},$$

the bound

$$T_2(\mathfrak{p}, V) \ll |V_\mathfrak{p}|^{5/2}$$

holds.

Proof Let $t = |V_{\mathfrak{p}}|$ and $n = (p-1)/t$. Let $a_1, \ldots, a_n$ be representatives of the distinct co-sets of the subgroup $V_{\mathfrak{p}}$ in $\Lambda_{\mathfrak{p}}^*$. We assume that they are ordered in such a way that

$$N_2(a_1, \mathfrak{p}, V) \geq N_2(a_2, \mathfrak{p}, V) \geq \cdots \geq N_2(a_n, \mathfrak{p}, V).$$

Without loss of generality, we may assume that t is large enough. Then for $1 \leq s < t^{1/2}$ and $B = \lfloor (2st)^{1/3} \rfloor + 1$, one verifies that

$$B < 2^{1/3}t^{1/2} + 1 \leq t, \qquad tB \leq t\left(2^{1/3}t^{1/2} + 1\right) < p, \qquad B^3 > 2ts.$$

Therefore the conditions of Lemma 3.2 are satisfied and it can be applied giving

$$\sum_{j=1}^{s} N_2(a_j, \mathfrak{p}, V) \ll s^{2/3}t^{2/3}.$$

Hence,

$$N_2(a_s, \mathfrak{p}, V) \ll s^{-1/3}t^{2/3}, \qquad 1 \leq s < t^{1/2}. \tag{3.13}$$

For $s \geq t^{1/2}$, the following estimate holds:

$$N_2(a_s, \mathfrak{p}, V) \leq N_2(a_{\lfloor t^{1/2} \rfloor}, \mathfrak{p}, V) \ll t^{1/2}. \tag{3.14}$$

Obviously, $N_2(0, \mathfrak{p}, V) \leq |V_{\mathfrak{p}}| = t$. Combining the bounds (3.13) and (3.14) with (3.2), we derive

$$
\begin{aligned}
T_2(\mathfrak{p}, V) &= \sum_{a \in \Lambda_{\mathfrak{p}}} N_2(a, \mathfrak{p}, V)^2 = \sum_{a \in \Lambda_{\mathfrak{p}}^*} N_2(a, \mathfrak{p}, V)^2 + N_2(0, \mathfrak{p}, V)^2 \\
&\leq t \sum_{s=1}^{n} N_2(a_s, \mathfrak{p}, V)^2 + t^2 \\
&\leq t \sum_{s \leq t^{1/2}} N_2(a_s, \mathfrak{p}, V)^2 + t \sum_{s > t^{1/2}} N_2(a_s, \mathfrak{p}, V)^2 + t^2 \\
&\ll t \sum_{s \leq t^{1/2}} \left(s^{-1/3}t^{2/3}\right)^2 + t \sum_{s > t^{1/2}} t^{1/2}N_2(a_s, \mathfrak{p}, V) + t^2 \ll t^{5/2},
\end{aligned}
$$

and have the desired result. $\qquad\qquad\square$

We use these inequalities to derive the upper bound on $S(\mathfrak{q}, V)$ which has been obtained in [33] and which improves several previously known bounds [44, 81, 82].

Theorem 3.4 *For any integer ideal* $\mathfrak{q}$ *which is relatively prime to each* $\lambda_1, \ldots, \lambda_r$,

$$S(\mathfrak{q}, V) \leq \mathrm{Nm}(\mathfrak{q})^{1/2}; \tag{3.15}$$

for any prime ideal $\mathfrak{p}$ *of first degree prime to each* $\lambda_1, \ldots, \lambda_r$,

$$S(\mathfrak{p}, V) \ll \mathrm{Nm}(\mathfrak{p})^{1/4} |V_\mathfrak{p}|^{3/8}; \tag{3.16}$$

and

$$S(\mathfrak{p}, V) \ll \mathrm{Nm}(\mathfrak{p})^{1/8} |V_\mathfrak{p}|^{5/8}. \tag{3.17}$$

Proof From (3.1), and the first inequality of Lemma 3.1 (with $k = 1$) we obtain (3.15).

We remark that now we may consider only the case

$$|V_\mathfrak{p}| < 0.7 \, \mathrm{Nm}(\mathfrak{p})^{2/3}$$

because for larger $|V_\mathfrak{p}|$, the bound (3.15) is better than both (3.16) and (3.17). From Lemma 3.3 (with $k = 2$) and the first and second inequalities of Lemma 3.1, we obtain (3.16) and (3.17), respectively.   □

We make several comments.

First of all, we note that we do not know if the bounds (3.16) and (3.17) are valid for arbitrary prime ideals. This is because the crucial point in the proof is Lemma 3.2. However, neither the method which is used here to prove this lemma nor the original method of proof of (3.8) from [17] can be extended to non-prime finite fields.

Question 3.5. *Improve the bound (3.15) for arbitrary integer ideals* $\mathfrak{q}$ *or at least for arbitrary prime ideals* $\mathfrak{p}$.

For some related estimates of exponential sums, see [45, 52, 84].

If $V_\mathfrak{q}$ is sufficiently large, the bound (3.15) is quite strong. In particular, it is non-trivial for $|V_\mathfrak{q}| > \mathrm{Nm}(\mathfrak{q})^{1/2}$. Unfortunately, we know no more than that, for any $\rho < (1 - \ln 2)/2 = 0.153 \ldots$,

$$|V_\mathfrak{p}| > \mathrm{Nm}(\mathfrak{p})^{r/(r+1)} \exp\left(\ln^\rho \mathrm{Nm}(\mathfrak{p})\right)$$

for some set of prime ideals $\mathfrak{p}$ of asymptotic density 1 (see Lemma 9.7 below, which is a generalization of the recent result of [72] concerning the case $\mathbb{K} = \mathbb{Q}$). For sets of almost all integer ideals, one obtains an even weaker result. The current best result is due to [61] where, for any $\tau < e^{-\gamma} = 0.561 \ldots$, where γ is the Euler constant, the bound

$$|V_\mathfrak{q}| \geq \mathrm{Nm}(\mathfrak{q})^{\tau r/(r+1)}$$

is stated for almost all integer ideals q. One sees that $\tau r/(r+1) > 1/2$ beginning with $r = 9$, thus (3.15) is applicable to groups with at least nine generators.

Clearly, if $r = 1$, then in order to get non-trivial results for some infinite sequence of ideals, we do need something essentially better than the 'square-root' bound (3.15). The bounds (3.16) and (3.17) provide such a result. In particular, the bound (3.17) is non-trivial for $|V_{\mathfrak{p}}| \geq c \operatorname{Nm}(\mathfrak{p})^{1/3}$ where $c > 0$ is some absolute constant.

The main advantage of the bound (3.15) is that it holds for arbitrary integer ideals.

Generally, it is useful to remark the following:

(i) The bound (3.15) supersedes (up to the value of constants) other estimates for

$$|V_{\mathfrak{p}}| \geq \operatorname{Nm}(\mathfrak{p})^{2/3}.$$

(ii) The bound (3.16) supersedes (up to the value of constants) other estimates for

$$\operatorname{Nm}(\mathfrak{p})^{2/3} \geq |V_{\mathfrak{p}}| \geq \operatorname{Nm}(\mathfrak{p})^{1/2}.$$

(iii) The bound (3.17) supersedes (up to the value of constants) other estimates for

$$\operatorname{Nm}(\mathfrak{p})^{1/2} \geq |V_{\mathfrak{p}}| \geq \operatorname{Nm}(\mathfrak{p})^{1/3}.$$

Obtaining explicit forms of the estimates (3.16) and (3.17) would be very important for a number of applications, including the question considered in Chapter 6 about the best value of the constant in the bound of Gaussian sums with arbitrary integer denominator.

Now we present the bounds (3.16) and (3.17) (as well as the bound (3.15), just for the sake of completeness) in a simple form as bounds on exponential sums with exponential functions. Let, as before, p be prime, g be an integer with $\gcd(g, p) = 1$, and t be the multiplicative order of g modulo p. Then the bound

$$\max_{\gcd(\alpha, p)=1} \left| \sum_{x=1}^{t} \mathbf{e}(\alpha g^x/p) \right| \ll \begin{cases} p^{1/2}, & \text{if } t \geq p^{2/3}; \\ p^{1/4}t^{3/8}, & \text{if } p^{1/2} \leq t \leq p^{2/3}; \\ p^{1/8}t^{5/8}, & \text{if } p^{1/3} \leq t \leq p^{1/2} \end{cases}$$

holds. Also, it can be useful to remark that Lemma 3.3 essentially claims that for $t \geq 0.7 p^{2/3}$ the bound

$$\sum_{\alpha=1}^{p-1} \left| \sum_{x=1}^{t} \mathbf{e}(\alpha g^x / p) \right|^4 \ll p t^{5/2}$$

holds.

Finally, we present the bounds (3.15), (3.16), and (3.17) in the form of bounds for Gaussian sums. Let p be a prime, then for any $n | p - 1$ and any g of multiplicative order $t = (p - 1)/n$, we have

$$\sum_{z=1}^{p-1} \mathbf{e}(\alpha z^n / p) = n \sum_{x=0}^{t-1} \mathbf{e}(\alpha g^x / p). \tag{3.18}$$

Denote

$$G_n(p) = \max_{\gcd(\alpha, p)=1} \left| \sum_{z=1}^{p} \mathbf{e}(\alpha z^n / p) \right|. \tag{3.19}$$

Then

$$G_n(p) \ll \begin{cases} n p^{1/2}, & \text{if } n \leq p^{1/3}; \\ n^{5/8} p^{5/8}, & \text{if } p^{1/3} \leq n \leq p^{1/2}; \\ n^{3/8} p^{3/4}, & \text{if } p^{1/2} \leq n \leq p^{2/3}. \end{cases} \tag{3.20}$$

We note that the first bound is a very well-known estimate, the other two estimates follow from the bounds (3.16) and (3.17), respectively.

Certainly, each estimate holds for all n; we have just pointed out when each one supersedes (up to the value of constants) the two others and remains non-trivial (one may easily check that the value $z = 0$ can be included in the sum without any harm).

If n is an even divisor of $p - 1$ such that $p \not\equiv 1 \pmod{2n}$, then paper [59] provides the explicit bound

$$G_n(p) \leq 2^{-1/2}(n^2 - 2n + 2)^{1/2} p^{1/2}$$

which improves the classical inequality $G_n(p) \leq (n-1)p^{1/2}$. This bound also immediately implies a similar improvement of (3.15) if $\mathfrak{q} = \mathfrak{p}$ is a prime ideal of first degree.

It is obvious that Lemma 3.3 implies that

$$T_3(\mathfrak{p}, V) \ll |V_{\mathfrak{p}}|^{9/2}. \tag{3.21}$$

Any stronger bound will immediately imply a non-trivial upper bound on $S(\mathfrak{p}, V)$ and $G_n(p)$ for a wider range of parameters.

Question 3.6. *Improve the bound (3.21).*

Now, in the case when q is a large power k of any unramified fixed prime ideal of first degree $\mathfrak{p}$, paper [44] provides the following much stronger estimate:

$$S(\mathfrak{p}^k, V) \leq c(\mathfrak{p})\,\mathrm{Nm}(\mathfrak{p}^k)|V_{\mathfrak{p}^k}|^{-1}, \qquad (3.22)$$

where $c(\mathfrak{p})$ is a constant depending upon $\mathfrak{p}$ only. In [78] this bound was generalized to character sums with linear recurrent sequences. That bound infers that, for any unramified prime ideal $\mathfrak{p}$ of an arbitrary degree d and for any $\varepsilon > 0$,

$$S(\mathfrak{p}^k, V) \leq c(\mathfrak{p}, \varepsilon)\,\mathrm{Nm}(\mathfrak{p}^k)|V_{\mathfrak{p}^k}|^{-1/d+\varepsilon}. \qquad (3.23)$$

We also mention the following nice estimate of [44]. Let $q = p^k$ be a power of a fixed prime number p and let g be an integer with $\gcd(p, g) = 1$. Denote by t the multiplicative order of g modulo q. Then, for any $P \leq t$, the bound

$$\sum_{x=1}^{P} \mathbf{e}(g^x/q) = O\left(P \exp\left(-\gamma \frac{\ln^3 P}{\ln^2 q}\right)\right)$$

holds, where $\gamma > 0$ is an absolute constant and the implied constant depends only on g and p. This bound is non-trivial even for very short sums of length P of order $\exp(\ln^{2/3+\varepsilon} q)$.

4

Bounds of Short Character Sums

Here we consider reasonably short character sums where the number of terms is small compared with the size of the corresponding algebraic domain.

First of all, we present the bound

$$G_n(p) \le p \left(1 - \frac{c(\varepsilon)}{\ln^{1+\varepsilon} p} \right), \tag{4.1}$$

which holds with some constant $c(\varepsilon) > 0$ provided that

$$p \ge \frac{n \ln n}{\mathrm{Ln}\, \ln^{1-\varepsilon} n}.$$

This bound is Theorem 1 of [41].

Therefore, for a prime ideal $\mathfrak{p}$ of first degree $\mathrm{Nm}(\mathfrak{p}) = p$ such that

$$|V_{\mathfrak{p}}| \ge \frac{\ln p}{\mathrm{Ln}\, \ln^{1-\varepsilon} p},$$

the bound

$$S(\mathfrak{p}, V) \le |V_{\mathfrak{p}}| \left(1 - \frac{C(\varepsilon)}{\ln^{1+\varepsilon} p} \right) \tag{4.2}$$

holds, where $C(\varepsilon)$ is a positive constant depending on an arbitrary $\varepsilon > 0$ only.

Moreover, from Lemma 6 of [41] one can easily derive the following estimate of an incomplete sum. Let t be the multiplicative order of g modulo p, $\gcd(g, p) = 1$ and let P be integer. If

$$t \ge \frac{\ln p}{\mathrm{Ln}\, \ln^{1-\varepsilon} p}, \qquad P \ge \min\{t,\, 6 \ln p \,\mathrm{Ln}\, \ln p + 1\}$$

then

$$\left| \sum_{x=1}^{P} \mathbf{e}(g^x/p) \right| \le P \left(1 - \frac{C(\varepsilon)}{\ln^{1+\varepsilon} p} \right).$$

26

We remark that we cannot obtain good upper estimates on $S(\mathfrak{p}, V)$ if $|V_{\mathfrak{p}}|$ is too small. It is known from [35] that, for any $\varepsilon > 0$, there are infinitely many primes p and subgroups $V_{\mathfrak{p}}$ of $\Lambda_{\mathfrak{p}}^{*}$ such that

$$\frac{\ln p}{\ln \varepsilon^{-1}} \leq V_{\mathfrak{p}} \leq \frac{2 \ln p}{\ln \varepsilon^{-1}}$$

and

$$S(\mathfrak{p}, V) \geq |V_{\mathfrak{p}}| \left(1 - \frac{2\pi\varepsilon}{1 - \varepsilon}\right).$$

Finally, we note that, over finite fields of bounded characteristics, the method of [81, 82] allows us to obtain a non-trivial estimate for fairly short sums. For example, if χ is a non-trivial additive character of $\mathbb{F}_{2^r}$ and $g \in \mathbb{F}_{2^r}^{*}$ is of multiplicative order $t \geq 2^{r\varepsilon}$, then, for any $a \in \mathbb{F}_{2^r}^{*}$ and any $\varepsilon > 0$, the bound

$$\left|\sum_{x=1}^{t} \chi(ag^{x})\right| \leq (1 - 2\gamma(\varepsilon))t + o(t)$$

holds, where $0 < \gamma(\varepsilon) < 1/2$ is the unique root of the equation

$$-\gamma \log \gamma - (1 - \gamma) \log(1 - \gamma) = 1 - \varepsilon, \qquad 0 < \gamma < 1/2.$$

More general estimates are obtained in [83].

In the next result, we demonstrate one more possible approach to upper bounds of such 'short' sums.

First of all, we need the following technical statement.

Lemma 4.1. *Let* $b_1, \ldots, b_t \in \mathbb{R}$ *be real numbers with* $|b_j| < p/2$, $j = 1, \ldots, t$, *which satisfy the inequality*

$$\sum_{j=1}^{t} b_j^2 \geq B.$$

Then

$$\Re \sum_{j=1}^{t} e(b_j/p) \leq t - \frac{8B}{p^2}.$$

Proof One easily verifies the inequality

$$\cos u \leq 1 - \frac{2u^2}{\pi^2}, \qquad -\pi \leq u \leq \pi.$$

Indeed, setting $f(u) = 1 - 2u^2/\pi^2 - \cos u$, we have $f(0) = f'(0) = f(\pi) = 0$. We also have $f(\pi/2) = 1/2 > 0$. If $f(x)$ takes a negative value on the interval $[-\pi, \pi]$, then, because it is an even function, it takes a negative value on the interval $[0, \pi]$ as well. Therefore f has at least four roots on the interval $[0, \pi]$ (counting the multiple root at $u = 0$ twice). Therefore, $f''(u) = -4/\pi^2 + \cos u$ has at least two zeros on the open interval $(0, \pi)$ which is not possible.

Therefore we have

$$\Re \sum_{j=1}^{t} e(b_j/p) = \sum_{j=1}^{t} \cos(2\pi b_j/p) \le t - \frac{8}{p^2} \sum_{j=1}^{t} b_j^2$$

and the assertion of the lemma follows.   □

We use Lemma 4.1 to prove the following upper bound.

Theorem 4.2 *For any integer r with $2 \le r \le \varphi(t)$, where t is the multiplicative order of g modulo p, the bound*

$$\max_{\gcd(a,p)=1} \left| \sum_{x=1}^{t} e(ag^x/p) \right| \le t - C(r)p^{-2/r}$$

holds, where $C(r) > 0$ depends only on r.

Proof Let us denote

$$B = \left(\frac{p^{2(r-1)}t}{r\gamma_{r-1}^{r-1}} \right)^{1/r}$$

where γ_k is the kth Hermite constant [10].

Let b be any integer with $\gcd(b, p) = 1$. For $j = 1, \ldots, t$ we define the numbers by the conditions

$$b_j \equiv bg^j \pmod{p}, \qquad |b_j| < p/2$$

and then we extend the definition to other j periodically with period t, that is we put $b_{t+1} = b_1, b_{t+2} = b_2$, and so on.

First of all, we show that

$$\sum_{j=1}^{t} b_j^2 \ge B. \tag{4.3}$$

Indeed, assuming the opposite we obtain that for some integer k,

$$\sum_{i=1}^{r} b_{k+i}^2 < rB/t.$$

Consider the $(r-1)$-dimensional lattice

$$L = \{(l_1, \ldots, l_r) \in \mathbb{Z}^r \ : \ b_{k+1}l_1 + \cdots + b_{k+r}l_r = 0\}.$$

The determinant of the lattice

$$\det L \leq \left(\sum_{i=1}^{r} b_{k+i}^2\right)^{1/2} < (rB/t)^{1/2}.$$

Therefore, from the definition of the Hermite constant, see [10], we derive that L contains a non-zero vector $(l_1, \ldots, l_r) \in L$ of length

$$\sum_{i=1}^{r} l_i^2 \leq \gamma_{r-1}(\det L)^{2/(r-1)} < \gamma_{r-1}(rB/t)^{1/(r-1)}.$$

For this vector (as well as for any other vector of L), we have

$$b_{k+1}l_1 + \cdots + b_{k+r}l_r = 0.$$

Hence

$$b_{j+1}l_1 + \cdots + b_{j+r}l_r \equiv 0 \quad (\mathrm{mod}\ p)$$

for any integer j. On the other hand,

$$|b_{j+1}l_1 + \cdots + b_{j+r}l_r| \leq \left(\sum_{i=1}^{r} b_{k+i}^2\right)^{1/2}\left(\sum_{i=1}^{r} l_i^2\right)^{1/2}$$
$$< B^{1/2}\gamma_{r-1}^{1/2}(rB/t)^{1/2(r-1)} = p.$$

Therefore, for all integers j, we have the identity

$$b_{j+1}l_1 + \cdots + b_{j+r}l_r = 0.$$

This means that the sequence (b_j) is a periodic linear recurrent sequence of order at most r with the smallest period t. Let $\Psi(X)$ be the minimal polynomial of this sequence. We have

$$\Psi(X) \mid l_r X^{r-1} + \cdots + l_2 X + l_1,$$

thus $\deg \Psi \leq r - 1$.

Because t is the smallest period then the smallest positive τ for which $\Psi(X) | X^\tau - 1$ is $\tau = t$. If

$$\Psi(X) \Big| \prod_{1 \leq \tau < t} (X^\tau - 1),$$

then $g^\tau \equiv 1 \ (\mathrm{mod}\ p)$ for some $1 \leq \tau < t$ which is impossible. Therefore one of the roots of Ψ is a primitive tth root of unity, in other words, Ψ is divisible by the t cyclotomic polynomial. Hence $\varphi(t) \leq r - 1$ which contradicts the assumption of the theorem. So the bound (4.3) holds.

Now we have,

$$
\left| \sum_{x=1}^{t} \mathbf{e}(ag^x/p) \right|^2 = \sum_{x,y=1}^{t} \mathbf{e}\left(a(g^x - g^y)/p\right) = \sum_{x,y=1}^{t} \mathbf{e}\left(a(g^y - 1)g^x/p\right)
$$

$$
= t + \sum_{y=1}^{t-1} \sum_{x=1}^{t} \mathbf{e}\left(a(g^y - 1)g^x/p\right)
$$

$$
= t + \sum_{y=1}^{t-1} \sum_{x=1}^{t} \Re\mathbf{e}\left(a(g^y - 1)g^x/p\right),
$$

because the sum on the left-hand side is a real number. If $\gcd(a, p) = 1$, then, for each $y = 1, \ldots, t - 1$, we can combine the bound (4.3) (with $b = a(g^y - 1)$) and Lemma 4.1 and obtain

$$
\left| \sum_{x=1}^{t} \mathbf{e}(ag^x/p) \right|^2 \leq t + (t-1)\left(t - \frac{8B}{p^2}\right) = t^2 - \frac{8B(t-1)}{p^2}
$$

and

$$
\left| \sum_{x=1}^{t} \mathbf{e}(ag^x/p) \right| \leq t - \frac{4(t-1)B}{tp^2}.
$$

Recalling the definition of B, we obtain the estimate

$$
\max_{\gcd(a,p)=1} \left| \sum_{x=1}^{t} \mathbf{e}(ag^x/p) \right| \leq t - C(r)t^{1/r}p^{-2/r}. \qquad (4.4)
$$

Replacing $t^{1/r}$ with 1, we obtain the desired result.   □

It is easy to see that, although the bound (4.4) looks stronger than the bound of Theorem 4.2, both statements are equivalent (up to the value of the constant $C(r)$).

Putting $r = \varphi(t)$, we obtain

$$
\max_{\gcd(a,p)=1} \left| \sum_{x=1}^{t} \mathbf{e}(ag^x/p) \right| \leq t - c(t)p^{-2/\varphi(t)} \qquad (4.5)
$$

where $c(t) > 0$ depends only on t.

We remark that there are close relations between bounds of such short sums and bounds of the number $M_1(m, p)$ which we define and study in Chapter 13. In particular, in that chapter we indicate another way of obtaining the bound (4.5) and also show that this bound is sharp.

5

Bounds of Character Sums for Almost All Moduli

So far we have considered estimates which hold for each q from some wide class of parameters. Below, we obtain a quite different result where we estimate character sums for almost all moduli (rather than for all) but instead we obtain a much stronger bound.

First of all, we need two auxiliary results which might be of independent interest. We consider relations of the form

$$\sum_{v=1}^{k} a_v \zeta_v = 0, \tag{5.1}$$

where the a_v are integers and the ζ_v are roots of unity. A relation (5.1) will be called irreducible if

$$\sum_{v \in \mathcal{I}} a_v \zeta_v \neq 0$$

for any non-empty proper subset $I \in \{1, \ldots, k\}$.

We will use the following statement from [56] (see also [77]).

Lemma 5.1. *If the relation (5.1) is irreducible, then there are distinct primes* $p_1, \ldots, p_s$ *where* $p_1 < \cdots < p_s \leq k$ *and* $p_1 \ldots p_s$th *roots of unity* η_v *such that*

$$\zeta_v = \eta_v \zeta, \qquad v = 1, \ldots, k$$

for some ζ.

Proof Let n_v be the order of ζ_v and let m be the least common multiple of n_v. Suppose that, for a prime p, we have $m = p^j r$, $j > 0$, $(p, r) = 1$. Then

$$\zeta_v = \rho^{\sigma_v} \xi_v, \qquad 0 \leq \sigma_v \leq p - 1,$$

31

where ρ is a primitive p^jth root of unity and ξ_ν a $p^{j-1}r$th root of unity. Collecting terms with the same value of σ_ν, we get from (5.1) the equation

$$\sum_{\mu=0}^{p-1} \rho^\mu \alpha_\mu = 0, \quad \alpha_\mu \in R(\xi), \tag{5.2}$$

where ξ is a primitive $p^{j-1}r$th root of unity. If $j > 1$, then ρ is of degree $\varphi(m)/\varphi(p^{j-1}r) = p$ over $R(\xi)$. Hence $\alpha_\mu = 0$ for all $\mu = 0, \ldots, p-1$. Because the relation (5.1) is irreducible, we conclude that, for some μ, we have

$$\sigma_1 = \cdots = \sigma_k = \mu.$$

Therefore,

$$\sum_{\nu=1}^{k} a_\nu \xi_\nu \neq 0 \tag{5.3}$$

and this relation is irreducible too.

If $j = 1$ and $p > k$, then all α_μ in (5.2) must be equal because

$$x^{p-1} + x^{p-2} + \cdots + 1 = 0$$

is the irreducible polynomial for ρ over $R(\xi)$. But (5.2) contains at most k non-vanishing terms. Hence we must have

$$\alpha_\mu = 0 \quad (\mu = 0, \ldots, p-1)$$

and again we obtain (5.3). Lemma 5.1 now follows by induction on m.   □

Now let t be an integer and $\xi = \mathbf{e}(1/t)$ be a tth root of unity. Denote by $W_k(t)$ the number of solutions of the equation

$$\xi^{z_1} + \cdots + \xi^{z_k} = \xi^{z_{k+1}} + \cdots + \xi^{z_{2k}}, \qquad 0 \le z_1, \ldots, z_{2k} \le t - 1. \tag{5.4}$$

Lemma 5.2. *For any fixed integer $k \ge 1$, the bound*

$$W_k(t) = A(k)t^k + O(t^{k-1})$$

holds, where

$$A(k) = \begin{cases} k!, & \text{if } t \text{ is odd}; \\ (2k-1)!!, & \text{if } t \text{ is even}. \end{cases}$$

Proof It follows from Lemma 5.1 that there is a constant $\gamma(k) > 0$ such that, for any integers $a_1, \ldots, a_m$, the equality

$$a_1 \xi^{z_1} + \cdots + a_m \xi^{z_m} = 0$$

with

$$\gcd(t, z_1 - z_m, \dots, z_{m-1} - z_m) < \gamma(k)t$$

implies that there is some proper vanishing subsum, that is for some proper subset $\mathcal{I} \in \{1, \dots, m\}$, we have

$$\sum_{i \in \mathcal{I}} a_i \xi^{z_i} = 0.$$

Namely, one can take

$$\gamma(k) = \prod_{\substack{p \in \mathcal{P} \\ p \le k}} p^{-1}$$

where $\mathcal{P}$ is the set of prime numbers.

Let us consider a partition

$$\mathcal{S} = \{\mathcal{I}_1, \dots, \mathcal{I}_s\}$$

of the set $\{1, \dots, 2k\}$ on s pairwise not intersecting subsets $\mathcal{I}_1, \dots, \mathcal{I}_s$, thus

$$\bigcup_{r=1}^{s} \mathcal{I}_r = \{1, \dots, 2k\}, \qquad \mathcal{I}_r \bigcap \mathcal{I}_l = \emptyset, \quad 1 \le l < r \le s.$$

We say that a solution of (5.4) is of type $\mathcal{S}$ if and only if

$$\sum_{i \in \mathcal{I}_r} \vartheta_i \xi^{z_i} = 0, \qquad r = 1, \dots, s,$$

where $\vartheta_i = 1$ if $i \le k$ and $\vartheta_i = -1$ otherwise, and no further subpartition satisfies this condition. In this case we see that, for any $r = 1, \dots, s$, numbers $z_i - z_j$, $i, j \in \mathcal{I}_r$ have some common divisor $d_r \ge \gamma(k)t$. Thus, for a given type $\mathcal{S} = \{\mathcal{I}_1, \dots, \mathcal{I}_s\}$ there are at most

$$\prod_{r=1}^{s} \left(t \sum_{\substack{d_r | t \\ d_r \ge \gamma(k)t}} (t/d_r)^{|\mathcal{I}_r|-1} \right) = \prod_{r=1}^{s} \left(t \sum_{\substack{\delta_r | t \\ \delta_r \le \gamma(k)^{-1}}} \delta_r^{|\mathcal{I}_r|-1} \right) \ll t^s$$

solutions. Therefore if $s \le k - 1$, then the number of solutions is $O(t^{k-1})$. The case $s = k$ means that we have a partition into pairs.

If t is odd, then the equation $\vartheta_i \xi^{z_i} + \vartheta_j \xi^{z_j} = 0$ is possible only if $\vartheta_i = -\vartheta_j$ and $z_i = z_j$. Therefore, the second half $(z_{k+1}, \dots, z_{2k})$ of each solution should be a permutation of the first half $(z_1, \dots, z_k)$. Thus if we fix the first half, then for the second half we have only $O(1)$ possibilities and precisely $k!$ possibilities if $0 \le z_1, \dots, z_k \le t - 1$ are pairwise distinct. The number of such pairwise distinct k-tuples is evidently $t^k + O(t^{k-1})$, the number of other k-tuples is $O(t^{k-1})$ and we get $k!t^k + O(t^{k-1})$ such solutions.

If t is even, then there is one more possibility: $\vartheta_i = \vartheta_j$ and $z_i = z_j + t/2$. Therefore, any partition of the set $\{1, \ldots, 2k\}$ on pairs is available. The total number of partitions is $(2k - 1)!!$. Similarly, in this case, we obtain $(2k - 1)!! t^k + O(t^{k-1})$ solutions. □

Lemma 5.3. *Let l be a prime number. Assume that $m \geq 1$ simple roots of the congruence $G(x) \equiv 0 \pmod{l}$ with a polynomial $G(X) \in \mathbb{Z}[X]$ are also roots of the congruence $F(x) \equiv 0 \pmod{l}$ with another polynomial $F(X) \in \mathbb{Z}[X]$. Then the resultant*

$$\mathrm{Res}(F, G) \equiv 0 \pmod{l^m}.$$

Proof Because each root α of G is simple,

$$G(\alpha) \equiv 0 \pmod{l}, \qquad G'(\alpha) \not\equiv 0 \pmod{l}.$$

By Theorem 3 of [39], α can be lifted to l-adic roots of $G(X)$. Therefore over the ring of integer l-adic integers $\mathbb{Z}_l$ we have $G = HQ$, where $H(X), Q(X) \in \mathbb{Z}_l[X]$, and H is the monic polynomial of degree m whose roots are l-adic liftings of those roots of G.

We write $F = HP + R$ where $P(X), R(X) \in \mathbb{Z}_l[X]$ and $\deg R < m$. From the condition of the lemma, we see that R has m roots modulo l, so it is identical to zero modulo l. Therefore, $R(X) = lT(X)$ where $T(X) \in \mathbb{Z}_l[X]$. Now we have

$$
\begin{aligned}
\mathrm{Res}(F, G) &= \mathrm{Res}(HP + lT, HQ) \\
&= \mathrm{Res}(HP + lT, H)\,\mathrm{Res}(HP + lT, Q) \\
&= \mathrm{Res}(lT, H)\,\mathrm{Res}(HP + lT, Q).
\end{aligned}
$$

Taking into account that $\mathrm{ord}_l \mathrm{Res}(lT, H) \geq m$, we obtain the statement. □

We believe that the restriction that the zeros of G are simple is redundant.

Question 5.4. *Prove an analogue of Lemma 5.3 for two polynomials $G(X)$, $F(X) \in \mathbb{Z}[X]$ having m common zeros (possibly multiple) modulo l.*

We note that the constant $\gamma(k)$ from the proof of Lemma 5.2 can be explicitly evaluated (see the comment at the end of [77]).

The implied constant in the following theorem may depend on k.

Theorem 5.5 *For each integer t and prime $l \equiv 1 \pmod{t}$, we fix some element $g_{t,l}$ of multiplicative order t modulo l. Then, for any fixed integer $k \geq 2$, and* *an arbitrary $U > 1$, the bound*

$$\max_{\gcd(\alpha, l)=1} \left| \sum_{x=0}^{t-1} e(\alpha g_{t,l}^x / l) \right| \ll t l^{1/2k^2} (t^{-1/k} + U^{-1/k^2})$$

holds for all primes $l \equiv 1 \pmod t$ except possibly at most $U / \ln U$ of them.

Proof We may assume that $k < (t-1)/2$. We estimate the number of solutions $T_k(l, t)$ of the congruence

$$g_{t,l}^{x_1} + \cdots + g_{t,l}^{x_k} \equiv g_{t,l}^{y_1} + \cdots + g_{t,l}^{y_k} \pmod l, \tag{5.5}$$

where

$$0 \le x_i, y_i \le t - 1, \qquad i = 1, \dots, k.$$

Therefore, for any $1 \le r \le t$, $\gcd(r, t) = 1$, $g_{t,l}^s$ is the root of the polynomial

$$F(X) = X^{u_1} + \cdots + X^{u_k} - X^{v_1} - \cdots - X^{v_k}, \tag{5.6}$$

where s, $1 \le s \le t - 1$, is defined from the congruence $sr \equiv 1 \pmod t$ and

$$u_i \equiv r x_i \pmod l, \qquad v_i \equiv r y_i \pmod l, \qquad i = 1, \dots, k.$$

Let $\Psi_t(X)$ be the tth cyclotomic polynomial. Since $g_{t,l}$ is of multiplicative order t modulo l then $\Psi_t(g_{t,l}^s) \equiv 0 \pmod l$. Therefore l divides the resultant $R_t(F)$ of F and Ψ_t.

We call a solution trivial if $R_t(F) = 0$ for at least one of the corresponding polynomials of the form (5.6) (and therefore for all $\varphi(t)$ such polynomials). Lemma 5.2 implies that each congruence (5.5) has $W_k(t) = O(t^k)$ trivial solutions.

Denote by Q the product

$$Q = \prod_F R_t(F)$$

over all $O(t^{2k})$ polynomials $F(X)$ of the form (5.6) such that

$$0 \le u_1, v_1, \dots, u_k, v_k \le t - 1, \quad \text{and} \quad R_t(F) \neq 0.$$

Let us fix W such that $1 \le W \le t^{2k}$. We denote by $\mathcal{L}_W$ the set of primes $l \equiv 1 \pmod t$ for which (5.5) has at least W non-trivial solutions. We see that for any $l \in \mathcal{L}_W$,

$$\mathrm{ord}_l Q \ge W \varphi(t). \tag{5.7}$$

Indeed, although some polynomial $F(X)$ of the form (5.6) may correspond to several, say m, distinct solutions of (5.5), in this case $F(X)$ and $\Psi_t(X)$ have at least m common roots and from Lemma 5.3 the inequality (5.7) follows anyway. Thus

$$t^{|\mathcal{L}_W|W\varphi(t)} \leq \prod_{l \in \mathcal{L}_W} l^{W\varphi(t)} \leq Q.$$

To estimate Q, we note that

$$|R_t(F)| = \prod_{\substack{j=1 \\ \gcd(j,t)=1}}^{t} |F(\mathbf{e}(j/t))| \leq (2k)^{\varphi(t)},$$

hence $\ln Q \ll \varphi(t)t^{2k}$. Therefore

$$|\mathcal{L}_W| \ll \frac{\ln Q}{W\varphi(t)\ln t} \leq\!\ll \frac{t^{2k}}{W \ln t}.$$

For a prime $l \notin \mathcal{L}_W$, we get $T_k(t,l) \leq W_k(t) + W$ thus the second bound of Lemma 3.1 implies

$$\max_{\gcd(\alpha,l)=1} \left| \sum_{x=0}^{r-1} \mathbf{e}(\alpha g_{t,l}^x/l) \right|$$

$$\ll l^{1/2k^2}(t^k + W)^{1/k^2}t^{1-2/k} = tl^{1/2k^2}(t^{-1/k} + W^{1/k^2}t^{-2/k}).$$

Selecting $W = \max(1, ct^{2k}/U)$ with some sufficiently large constant c provided that $U > c$, we obtain the assertion. If $U \leq c$, then the required inequality holds for all l, without any exceptions. $\qquad\square$

Here is an example of using Theorem 5.5. Let t and A be such that there are at least $t^{A-1-\varepsilon}$ primes $l \equiv 1 \pmod{t}$ and such that $t^A \leq l \leq 2t^A$. Choosing $U = t^{A-1-\varepsilon k^2}$ we obtain that, for almost all primes l described above, the bound

$$\max_{\gcd(\alpha,l)=1} \left| \sum_{x=0}^{t-1} \mathbf{e}(\alpha g_{t,l}^x/l) \right| \ll t(t^{-(2k-A)/2k^2} + t^{(-A+2)/2k^2+\varepsilon})$$

holds. It is easy to see that the bound is non-trivial if $2 < A < 2k$. (For $A \leq 2$ one can use (3.16) or (3.17) both of which are non-trivial for such A.) If A is an integer, then the optimal choice for k is $k = A - 1$ which produces the estimate

$$\max_{\gcd(\alpha,l)=1} \left| \sum_{x=0}^{t-1} \mathbf{e}(\alpha g_{t,l}^x/l) \right| \ll t^{1-(A-2)/2(A-1)^2+\varepsilon}$$

for almost all l satisfying the conditions above. Another example of using Theorem 5.5 is given in the proof of Theorem 14.5 below.

Using (3.18) we can apply it to Gaussian sums as well.

6

Bounds of Gaussian Sums

For integers $q \geq 1$, $n \geq 2$, and a, let us denote by $S_n(a, q)$ the Gaussian sum

$$S_n(a, q) = \sum_{x=1}^{q} \mathbf{e}(ax^n/q)$$

and let

$$G_n(q) = \max_{\gcd(a,q)=1} |S_n(a, q)|$$

be the maximal absolute value of 'non-trivial' Gaussian sums with the denominator q.

In particular, for a prime $q = p$, this function has already been defined in Chapter 3. Accordingly, the bound (3.20) is one of our main tools.

It is very well known that for some constant $C(n)$ depending only n the bound

$$G_n(q) \leq C(n) \, q^{1-1/n}$$

holds (see [86] or Problem 3δ of Chapter 7 of [91] for example). So, we can define

$$A(n) = \sup_{q \geq 1} \frac{G_n(q)}{q^{1-1/n}}.$$

In [86], the bound

$$A(n) = \exp\left(O(\ln \ln n)^2\right)$$

was proved and it was also conjectured in that paper that actually all these constants are uniformly bounded, that is

$$A(n) = O(1).$$

37

In [81, 82], this conjecture was proved in the following stronger form:

$$A(n) = 1 + O(n^{-1/4+\varepsilon}).$$

Here we use the results of (4.2).

If $n = 2$, it is known that $A(2) = 2^{1/2}$. So, hereafter we consider the case of $n > 2$.

First of all, we recall some well-known properties of Gaussian sums. These properties are summarized in Section 2 of [86] (see Problem 12d of Chapter 3 and Problems 3β and 3γ of Chapter 7 of [91] as well).

Lemma 6.1. *For any relatively prime q_1, q_2, the identity*

$$G_n(q_1 q_2) = G_n(q_1) G_n(q_2)$$

holds.

Proof Let $q = q_1 q_2$. There is a one-to-one corresponding assigning to any integer $x \in \{1, \dots, q\}$ the integers $x_1 \in \{1, \dots, q_1\}$ and $x_2 \in \{1, \dots, q_2\}$ such that $x \equiv x_1 q_2 + x_2 q_1 \pmod{q}$. Therefore,

$$S_n(a, q) = \sum_{x_1=1}^{q_1} \sum_{x_2=1}^{q_2} \mathbf{e}(a(x_1 q_2 + x_2 q_1)^n / q).$$

Note that $(x_1 q_2 + x_2 q_1)^n \equiv (x_1 q_2)^n + (x_2 q_1)^n \pmod{q}$. Hence,

$$
\begin{aligned}
S_n(a, q) &= \sum_{x_1=1}^{q_1} \sum_{x_2=1}^{q_2} \mathbf{e}(a((x_1 q_2)^n + (x_2 q_1)^n)/q) \\
&= \sum_{x_1=1}^{q_1} \sum_{x_2=1}^{q_2} \mathbf{e}(a q_2^{n-1} x_1^n / q_1 + a q_1^{n-1} x_2^n / q_2) \\
&= \sum_{x_1=1}^{q_1} \mathbf{e}(a q_2^{n-1} x_1^n / q) \sum_{x_2=1}^{q_2} \mathbf{e}(a q_1^{n-1} x_2^n / q),
\end{aligned}
$$

and we have

$$S_n(a, q) = S_{n_1}(a_1, q_1) S_{n_2}(a_2, q_2), \tag{6.1}$$

where $a_1 = a q_2^{n-1}$, $a_2 = a q_1^{n-1}$. The equality (6.1) immediately gives $G_n(q) \leq G_n(q_1) G_n(q_2)$. The inverse inequality $G_n(q) \geq G_n(q_1) G_n(q_2)$ follows from the observation that for any $(a_1, q_1) = 1$ and $(a_2, q_2) = 1$, there exists a such that $a_1 \equiv a q_2^{n-1} \pmod{q_1}$ and $a_2 \equiv a q_1^{n-1} \pmod{q_2}$. $\qquad \square$

To prove Lemmas 6.4 and 6.5 below, we need some auxiliary assertions.

Lemma 6.2. *Let p be a prime and let $n \geq 1$, $\alpha \geq 0$, $\beta \geq 1$, x, y be integers such that $n = p^\alpha n_0$ with $\gcd(n_0, p) = 1$ and $y \equiv x \pmod{p^\beta}$.*

- *If $\beta \geq 1$, then*
$$y^n - x^n \equiv nx^{n-1}(y - x) \pmod{p^{\beta+1}};$$

- *If $(x, p) = 1$, $\beta \geq 2$, then*
$$y^n - x^n \equiv n_0 x^{n_0-1}(y - x)p^\alpha \pmod{p^{\alpha+\beta+1}}.$$

Proof We have

$$y^n - x^n - nx^{n-1}(y - x) = (y - x)\left(\sum_{j=0}^{n-1} x^j y^{n-1-j} - nx^{n-1}\right)$$

$$= (y - x)\left(\sum_{j=0}^{n-1} x^j (y^{n-1-j} - x^{n-1-j})\right) \equiv 0 \pmod{p^{\beta+1}},$$

as $y - x$ is divisible by p^β and all differences $y^{n-1-j} - x^{n-1-j}$ are divisible by p, and the first statement follows.

We prove the second statement by induction on α. For $\alpha = 0$, the assertion has just been proved. Denote $m = n/p$, $u = x^m$, $v = y^m$ and make the induction assumption that

$$v - u \equiv n_0 x^{n_0-1}(y - x)p^{\alpha-1} \pmod{p^{\alpha+\beta}}. \tag{6.2}$$

The last congruence implies that

$$v - u \equiv 0 \pmod{p^{\alpha+\beta-1}}.$$

Proceeding as in the first part, we obtain

$$v^p - u^p - p(v - u) = (v - u)\left(\sum_{j=0}^{p-1} u^j v^{p-1-j} - p\right). \tag{6.3}$$

From (6.2) we see that $u \equiv v \pmod{p^2}$. Therefore,

$$\sum_{j=0}^{p-1} u^j v^{p-1-j} - p \equiv \sum_{j=0}^{p-1} u^{p-1} - p = pu^{p-1} - p \equiv 0 \pmod{p^2},$$

and, by (6.3),

$$y^n - x^n = v^p - u^p \equiv p(v - u) \pmod{p^{\alpha+\beta+1}}.$$

Substituting (6.2) into the last congruence, we complete the proof. $\square$

Lemma 6.3. *Let p be a prime. We put*

$$\gamma_p(n) = \begin{cases} 2, & \text{if ord}_p\, n = 0; \\ 3 + \text{ord}_p\, n, & \text{if ord}_p\, n > 0. \end{cases}$$

Then, for any integer m with $\gamma_p(n) \le m \le n$,

$$\sum_{\substack{1 \le x \le p^m \\ (x,p)=1}} \mathbf{e}(ax^n/p^m) = 0.$$

Proof Let

$$S = \sum_{\substack{1 \le x \le p^m \\ (x,p)=1}} \mathbf{e}(ax^n/p^m).$$

We will use the following fact: for any integer y

$$S = \sum_{\substack{1 \le x \le p^m \\ (x,p)=1}} \mathbf{e}(a(x+py)^n/p^m)$$

(because $x + py$ runs over the reduced residue system modulo p^m together with x). Therefore, for any integer $\beta \ge 1$, we have

$$S = \frac{1}{p} \sum_{\substack{1 \le x \le p^m \\ (x,p)=1}} \sum_{j=0}^{p-1} \mathbf{e}(a(x+jp^\beta)^n/p^m). \tag{6.4}$$

Now set $\alpha = \text{ord}_p\, n$, $n_0 = n/p^\alpha$, $\beta = m - \alpha - 1$.
If $\alpha = 0$, then $\beta = m - 1 \ge \gamma_p(n) - 1 = 1$, and, by Lemma 6.2, we have

$$a(x+jp^\beta)^n - ax^n \equiv anjx^{n-1}p^{m-1} \pmod{p^m}.$$

Hence,

$$\sum_{j=0}^{p-1} \mathbf{e}(a(x+jp^\beta)^n/p^m) = \mathbf{e}(ax^n/p^m) \sum_{j=0}^{p-1} \mathbf{e}(anjx^{n-1}/p) = 0$$

and by (6.4) we have $S = 0$.
Similarly, if $\alpha > 0$, then $\beta = m - \alpha - 1 \ge \gamma_p(n) - \text{ord}_p\, n - 1 \ge 2$, and, by Lemma 6.2, we have

$$a(x+jp^\beta)^n - ax^n \equiv an_0jx^{n_0-1}p^{\alpha+\beta} \pmod{p^m}.$$

Hence,

$$\sum_{j=0}^{p-1} \mathbf{e}(a(x+jp^\beta)^n/p^m) = \mathbf{e}(ax^n/p^m) \sum_{j=0}^{p-1} \mathbf{e}(an_0jx^{n_0-1}/p) = 0$$

and by (6.4) we have $S = 0$ again. $\qquad\square$

Now we are prepared to prove the following two assertions giving a recurrent formula and initial values for Gaussian sums.

Lemma 6.4. *Let p be a prime. We put*

$$\gamma_p(n) = \begin{cases} 2, & \text{if } \operatorname{ord}_p n = 0; \\ 3 + \operatorname{ord}_p n, & \text{if } \operatorname{ord}_p n > 0. \end{cases}$$

Then, for any integer m with $\gamma_p(n) \le m \le n$,

$$G_n(p^m) = p^{m-1}.$$

Proof By Lemma 6.3,

$$\begin{aligned} S_n(a, p^m) &= \sum_{\substack{1 \le x \le p^m \\ (x,p)>1}} \mathbf{e}(ax^n/p^m) = \sum_{y=1}^{p^{m-1}} \mathbf{e}(a(py)^n/p^m) \\ &= \sum_{y=1}^{p^{m-1}} \mathbf{e}(ay^n p^{n-m}) = p^{m-1}, \end{aligned}$$

and we are done. □

Lemma 6.5. *Let p be a prime and let $m \ge n+1$ be an integer, then*

$$G_n(p^m) = p^{n-1} G_n(p^{m-n}).$$

Proof Because $n \ge 3$, we have $n \ge \operatorname{ord}_p n + 2$. Indeed, it is easy to verify that, for $x \ge 4$,

$$x \ge \log x + 2 \ge \operatorname{ord}_p x + 2$$

and also that $3 \ge \operatorname{ord}_p 3 + 2$. Therefore $m \ge n+1 \ge \operatorname{ord}_p n + 3 \ge \gamma_p(n)$. By Lemma 6.3, we obtain

$$S_n(a, p^m) = \sum_{y=1}^{p^{m-1}} \mathbf{e}(a(py)^n/p^m) = \sum_{y=1}^{p^{m-1}} \mathbf{e}(ay^n/p^{m-n}) = p^{n-1} S_n(a, p^{m-n}),$$

and we are done. □

Lemma 6.1 shows that it is enough to consider Gaussian sums with a prime power denominator, that is

$$A(n) = \prod_{p \in \mathcal{P}} \sup_{k \ge 1} \{ G_n(p^k)/p^{k(1-1/n)}, 1 \}.$$

Lemma 6.5 shows that

$$\sup_{k \geq 1}\{G_n(p^k)/p^{k(1-1/n)}, 1\} = \max_{n \geq m \geq 1}\{G_n(p^m)/p^{m(1-1/n)}, 1\}.$$

Finally, Lemma 6.4 shows that

$$\max_{n \geq m \geq 1}\{G_n(p^m)/p^{m(1-1/n)}, 1\} = \max_{\gamma_p(n) > m \geq 1}\{G_n(p^m)/p^{m(1-1/n)}, 1\}.$$

Therefore

$$A(n) = \prod_{p \in \mathcal{P}} \max_{\gamma_p(n) > m \geq 1}\{G_n(p^m)/p^{m(1-1/n)}, 1\},$$

where $\mathcal{P}$ denotes the set of prime numbers.

We estimate the part of the product which is taken over primes $p \mid n$ by using the following trivial inequalities:

$$\prod_{\substack{p \in \mathcal{P} \\ p \mid n}} \max_{\gamma_p(n) > m \geq 1}\{G_n(p^m)/p^{m(1-1/n)}, 1\} \leq \prod_{\substack{p \in \mathcal{P} \\ p \mid n}} \max_{\gamma_p(n) > m \geq 1}\{p^{m/n}, 1\}$$

$$\leq \prod_{\substack{p \in \mathcal{P} \\ p \mid n}} p^{(\gamma_p(n)-1)/n}$$

$$= n^{1/n} \prod_{\substack{p \in \mathcal{P} \\ p \mid n}} p^{2/n} \leq n^{3/n},$$

thus

$$A(n) \leq n^{3/n} \prod_{\substack{p \in \mathcal{P} \\ \gcd(p,n)=1}} \max\{G_n(p)/p^{1-1/n}, 1\}.$$

Now we are going to show that in fact we may consider only a finite product.

Lemma 6.6. *Let $d = \gcd(n, p-1)$, then $G_n(p) = G_d(p)$.*

Proof Evidently, for any l with $\gcd(l, p-1) = 1$, the function $f(y) = y^l$ is a permutation function on the set or residues modulo p. □

Let us define $Q_n(d)$ as the largest prime p with $\gcd(n, p-1) = d$ and such that $G_n(p) \geq p^{1-1/n}$. Therefore, the inequality

$$A(n) \leq n^{3/n} \prod_{d \mid n} \prod_{\substack{p \in \mathcal{P}, \ p \leq Q_n(d) \\ \gcd(n,p-1)=d}} \max\{G_d(p)/p^{1-1/n}, 1\} \qquad (6.5)$$

holds. Then the bound (3.20) implies that

$$Q_n(d) \ll d^{3/2}. \qquad (6.6)$$

Certainly the implied constant (as well as the constants in (3.20)) can be easily evaluated explicitly.

Theorem 6.7 *The asymptotic formula*

$$A(n) = 1 + O(n^{-1}\tau(n)\ln n)$$

holds.

Proof Let $p \leq Q_n(d)$ be a prime with $\gcd(p-1, n) = d \geq 2$. We put $t = (p-1)/d$ and show that if $\varphi(t) \geq 6$, then $\sigma_d(p) \leq p^{1-1/n}$ provided that n is large enough.

First of all, we remark that (6.6) implies that $d \gg p^{2/3}$, thus $t = O(p^{1/3})$.

Now, from the identity (3.18) and the estimate (4.4), taken with $r = 6$, we obtain

$$\begin{aligned}\sigma_d(p) &\leq 1 + d(t - Ct^{1/6}p^{-1/3}) = p - Cdt^{1/6}p^{-1/3}\\ &= p - C(p-1)t^{-5/6}p^{-1/3} \leq p - cp^{7/18}\end{aligned}$$

for some absolute constants $C, c > 0$.

Noticing that $7/18 > 1/3$, we see that there exists an absolute constant L_0 such that if $p \geq L_0$, then

$$p - cp^{7/18} \leq p - d^{-1}p\ln p \leq p^{1-1/d} \leq p^{1-1/n}.$$

Clearly, for $p < L_0$ and sufficiently large n, we also have $\sigma_d(p) \leq p^{1-1/n}$.

Now, by (6.5) we can write for sufficiently large n

$$\begin{aligned}A(n) &\leq n^{3/n} \prod_{d|n} \prod_{\substack{p\in\mathcal{P}\\ \gcd(n,p-1)=d\\ \varphi((p-1)/d)\leq 5}} \max\{\sigma_d(p)/p^{1/n}, 1\}\\ &\leq n^{3/n} \prod_{d|n} \prod_{\substack{p\in\mathcal{P}\\ \gcd(n,p-1)=d\\ \varphi((p-1)/d)\leq 5}} p^{1/n}\\ &\leq n^{3/n} \prod_{\varphi(t)\leq 5} \prod_{d|n}(td+1)^{1/n} = n^{3/n} \prod_{t\in\mathcal{T}} \prod_{d|n}(td+1)^{1/n}\end{aligned}$$

where $\mathcal{T}$ is the following nine-element set

$$\mathcal{T} = \{1, 2, 3, 4, 5, 6, 8, 10, 12\}.$$

Therefore

$$A(n) \leq n^{3/n}(12n+1)^{9\tau(n)/n} \leq 1 + O(\tau(n)\ln n/n),$$

as required. $\square$

In particular, one obtains that

$$A(n) \le 1 + n^{-1} \exp\left((\ln 2 + o(1)) \frac{\ln n}{\ln \ln n}\right), \qquad n \to \infty. \qquad (6.7)$$

Below, we prove a lower bound on $A(n)$ which shows that the previous estimate (6.7) is quite precise.

Theorem 6.8 *There exist infinitely many integers n such that the bound*

$$A(n) > 1 + n^{-1} \exp(0.43 \ln n / \ln \ln n)$$

holds.

Proof Let us fix arbitrary $A > 12/5$ and let $2 \le u < A + 1$. We also fix sufficiently small $\varepsilon > 0$, in particular we assume that $\varepsilon < (A + 1 - u)/5$. Let z be sufficiently large and let $x = z^A$.

From Theorem 2.1 of [1], we derive that there are primes $q_1, \ldots, q_k$, $k \le k_0(A)$, where $k_0(A)$ depends only on A, such that for any positive integer $d \le z$, we have

$$\pi(dx/z; d, 1) - \pi(dx/(2z); d, 1) > \frac{dx}{4\varphi(d)z \ln x} \ge \frac{x}{4z \ln x}, \qquad (6.8)$$

provided that d is not divisible by $q_1, \ldots, q_k$.
Let Q be the set of the least

$$R = \left\lfloor \frac{(1 - \varepsilon)u \ln z}{\ln \ln z} \right\rfloor$$

primes distinct from $q_1, \ldots, q_k$. Denote by N their product

$$N = \prod_{q \in Q} q.$$

We have

$$N < z^u \qquad (6.9)$$

provided that z is large enough.
Let D be the set of all products of

$$r = \left\lfloor \frac{(1 - \varepsilon) \ln z}{\ln \ln z} \right\rfloor$$

elements of Q, then for any $d \in D$, we have

$$z^{1-2\varepsilon} < d < z. \qquad (6.10)$$

Denote by $M = |D|$ the cardinality of D. Then

$$M = \binom{R}{r} > (u^u/(u-1)^{u-1})^{(1-2\varepsilon)\ln z/\ln\ln z}, \qquad (6.11)$$

see [54], Section 10.11.

Using (6.8), we establish the existence of l with $x/2z < l \le x/z$ and such that the numbers $ld + 1$ are prime for at least $M/(4\ln x)$ values of d. Set $n = lN$. By (6.9), we have

$$n < z^{u+A-1}. \qquad (6.12)$$

Because $\varepsilon < (A + 1 - u)/5$, we see from the previous inequality and from (6.10) that $ld + 1 > xz^{-2\varepsilon}/2 > n^{1/2}$ for any $d \in D$. We also note that $ld \mid n$ for any $d \in D$. Therefore for any prime $p = ld + 1, d \in D$, we have

$$
\begin{aligned}
\sigma_n(p) &\ge |S_n(1, p)| = |(p - 1)\mathbf{e}(1/p) + 1| = |(p - 1) + \mathbf{e}(-1/p)| \\
&\ge \Re\left((p - 1) + \mathbf{e}(-1/p)\right) = p - 1 + \cos(2\pi/p) \\
&= p + O(p^{-2}) = p + O(1/n) \ge p^{1-1/n}\exp(1/n),
\end{aligned}
$$

provided that z is large enough.
Therefore,

$$A(n) \ge \exp(M/(4n\ln x)). \qquad (6.13)$$

If we take $u = 3$, then

$$(u\ln u - (u - 1)\ln(u - 1))/(u + 7/5) > 0.43.$$

Now if we select A sufficiently close to $12/5$, then from the last inequality, and the inequalities (6.11), (6.12), and (6.13) we obtain the required estimate. $\square$

Certainly the constant 0.43 can be slightly improved. Also, it is useful to recall that $\ln 2 = 0.693\ldots$, thus the gap between the upper bound (6.7) and the lower bound of Theorem 6.8 is quite narrow.

It is important to remark that the implied constants in the above bounds of $A(n)$ can be explicitly evaluated. Thus with some reasonable amount of computation, one can perhaps obtain an absolute upper bound on all $A(n)$, $n = 2, 3, \ldots$. The first step would be to evaluate explicitly and to get the best possible values of the constants in the estimates (3.20) and (6.6).

In fact we believe that even the following question is quite feasible and can be answered within a reasonable amount of computation.

Question 6.9. *Compute the precise value of*

$$A = \sup_{n \ge 2} A(n).$$

Part three

Multiplicative Translations of Sets

7

Multiplicative Translations of Subgroups of $\mathbb{F}_p^*$

Let V be a subgroup of a prime field of p elements $\mathbb{F}_p^*$ of cardinality $|V| = t$.

We put $n = (p-1)/t$ and for $j = 1, \ldots, n$, we denote by V_j the multiplicative translation of V by g^j where g is a fixed primitive root modulo p, that is

$$V_j = g^j V = \left\{ g^j u \ : \ u \in V \right\}, \qquad j = 1, \ldots, n.$$

As we mentioned in the Introduction, many of the results of this book depend on good upper bounds on the size of the largest interval which does not contain elements of V_j for some $j = 1, \ldots, n$. Thus we define

$$
\begin{aligned}
H_p(t) \quad = \quad \max\{H \ : \ \exists j \in \{1, \ldots, n\}, \ \exists M \in \mathbb{Z} \\
\text{such that } M + z \notin V_j, \ z = 1, \ldots, H\}.
\end{aligned}
$$

To study $H_p(t)$, we introduce several more functions.

Namely, for $j = 1, \ldots, n$, we denote by $N_{j,t}(h)$ the number of $u \in V_j$ with $1 \le |u| \le h$, that is

$$N_{j,t}(h) = \left| \{ u \in V_j \ : \ 1 \le |u| \le h \} \right|,$$

and also put

$$S_j(t) = \sum_{v \in V_j} \mathbf{e}(v/p)$$

and extend this definition for all $j \in \mathbb{Z}$ periodically with period n.

Relations between $H_p(t)$ and $N_{j,t}(h)$, $S_j(t)$, are given by the following statement.

Lemma 7.1. *If for some integer $h \ge 1$, the inequality*

$$\sum_{j=1}^{n} N_{j,t}(h) |S_{j+k}(t)| \le 0.5t$$

49

is satisfied for all $k = 1, \ldots, n$, *then for any* $\varepsilon > 0$, *the bound*

$$H_p(t) \ll p^{1+\varepsilon} h^{-1}$$

holds.

Proof Let us fix some $\varepsilon > 0$. We put

$$s = \left\lceil 0.5(1 + \varepsilon^{-1}) \right\rceil, \qquad L = \left\lceil p^{1+\varepsilon} h^{-1} \right\rceil.$$

Obviously it is enough to show that for any integer M and any $k = 1, \ldots, n$, the congruence

$$
\begin{aligned}
v &\equiv M + x_1 + \cdots + x_s - y_1 - \cdots - y_s \pmod{p}, \\
v &\in V_k, \quad 0 \le x_1, y_1, \ldots, x_s, y_s < L,
\end{aligned}
$$

is solvable. Indeed, in this case we have $H_p(t) \le 2s(L - 1)$.

For the number Q of solutions of this congruence, one easily sees from the identity (3.3) applied to $\mathbb{F}_p$ that

$$
\begin{aligned}
Q &= \frac{1}{p} \sum_{a=-(p-1)/2}^{(p-1)/2} \mathbf{e}(-aM/p) \sum_{v \in V_k} \mathbf{e}(av/p) \\
&\quad \times \sum_{0 \le x_1, y_1, \ldots, x_s, y_s < L} \mathbf{e}\left(-\frac{a(x_1 + \cdots + x_s - y_1 - \cdots - y_s)}{p} \right) \\
&= \frac{1}{p} \sum_{a=-(p-1)/2}^{(p-1)/2} \mathbf{e}(-aM/p) \left| \sum_{0 \le x < L} \mathbf{e}(ax/p) \right|^{2s} \sum_{v \in V_k} \mathbf{e}(av/p) \\
&\ge tL^{2s} p^{-1} - \sigma_1 p^{-1} - \sigma_2 p^{-1},
\end{aligned}
$$

where

$$\sigma_1 = \sum_{1 \le |a| \le h} \left| \sum_{0 \le x < L} \mathbf{e}(ax/p) \right|^{2s} \left| \sum_{v \in V_k} \mathbf{e}(av/p) \right|,$$

$$\sigma_2 = \sum_{h < |a| \le (p-1)/2} \left| \sum_{0 \le x < L} \mathbf{e}(ax/p) \right|^{2s} \left| \sum_{v \in V_k} \mathbf{e}(av/p) \right|.$$

For $1 \le |a| \le h$, we use the trivial estimate

$$\left| \sum_{0 \le x < L} \mathbf{e}(ax/p) \right| \le L.$$

and derive

$$\sigma_1 \leq L^{2s} \sum_{1 \leq |a| \leq h} \left| \sum_{v \in V_k} e(av/p) \right| = L^{2s} \sum_{j=1}^{n} \sum_{\substack{1 \leq |a| \leq h \\ a \in V_j}} \left| \sum_{v \in V_k} e(av/p) \right|$$

$$= L^{2s} \sum_{j=1}^{n} \sum_{\substack{1 \leq |a| \leq h \\ a \in V_j}} \left| \sum_{v \in V_{j+k}} e(v/p) \right| = L^{2s} \sum_{j=1}^{n} N_{j,t}(h) |S_{j+k}(t)|$$

$$\leq 0.5 t L^{2s}.$$

If $h < |a| \leq (p-1)/2$, then from (3.5) we see

$$\left| \sum_{0 \leq x < L} e(ax/p) \right| \ll p/h \ll Lp^{-\varepsilon}.$$

And because trivially

$$\left| \sum_{v \in V_k} e(av/p) \right| \leq t,$$

we obtain

$$\sigma_2 \ll \sum_{h < |a| \leq (p-1)/2} t L^{2s} p^{-2s\varepsilon} \leq t L^{2s} p^{1-2s\varepsilon} \ll t L^{2s} p^{-\varepsilon}.$$

Therefore $Q \geq 0.5 t L^{2s} p^{-1} + O(t L^{2s} p^{-1-\varepsilon})$, thus $Q > 0$ provided that p is large enough and the claimed result follows. $\qquad\square$

For a non-negative integer $h < p$, we denote by $N_t(h)$ the number of solutions of the congruence

$$ux \equiv y \pmod{p}, \qquad 0 < |x|, |y| \leq h, \ u \in V.$$

Thus for any α with $1/4 \leq \alpha \leq 1/2$ and any $k = 1, \ldots, n$ we have

$$\sum_{j=1}^{n} N_{j,t}(h) |S_{j+k}(t)| \leq \left(\sum_{j=1}^{n} N_{j,t}(h)^2 \right)^{\alpha} \left(\sum_{j=1}^{n} N_{j,t}(h) \right)^{1-2\alpha}$$

$$\times \left(\sum_{j=1}^{n} |S_j(t)|^2 \right)^{2\alpha - 1/2} \left(\sum_{j=1}^{n} |S_j(t)|^4 \right)^{1/2 - \alpha}.$$

Clearly,

$$\sum_{j=1}^{n} N_{j,t}(h)^2 = N_t(h), \qquad \sum_{j=1}^{n} N_{j,t}(h) = 2h. \tag{7.1}$$

The other two sums can be expressed in term of the quantities $\sigma_{2s}(p, V)$ and $T_s(p, V)$ which have been defined in Chapter 3.

$$\sum_{j=1}^{n} |S_j(t)|^{2s} < \frac{1}{t}\sigma_{2s}(p, V) = \frac{p}{t}T_s(p, V).$$

Therefore we see that for any α with $1/4 \le \alpha \le 1/2$ and any $k = 1, \ldots, n$, we have

$$\sum_{j=1}^{n} N_{j,t}(h)|S_{j+k}(t)| \le (2h)^{1-2\alpha} p^\alpha t^{\alpha-1/2} N_t(h)^\alpha T_4(p, V)^{1/2-\alpha}. \qquad (7.2)$$

In particular, for any t, we can select $\alpha = 1/2$ and get the inequality

$$\sum_{j=1}^{n} N_{j,t}(h)|S_{j+k}(t)| \ll p^{1/2}N_t(h)^{1/2}. \qquad (7.3)$$

For $t < 0.7p^{2/3}$, Lemma 3.3 provides an upper bound on $T_4(p, V)$ from which, selecting $\alpha = 1/4$, we derive

$$\sum_{j=1}^{n} N_{j,t}(h)|S_{j+k}(t)| \ll p^{1/4}h^{1/2}N_t(h)^{1/4}t^{3/8}. \qquad (7.4)$$

Now our immediate goal is to get good upper bounds on $N_t(h)$.

Obviously, the function $N_t(h)$ is non-decreasing. Also for $0 \le h < p$ the bounds

$$4h^2 \ge N_t(h) \ge \max\left\{2h, \frac{4h^2}{n}\right\} = \max\left\{2h, \frac{4h^2t}{p-1}\right\} \qquad (7.5)$$

hold. Indeed, the upper bound and the bound $N_t(h) \ge 2h$ are trivial, and from the representations (7.1), we derive

$$4h^2 = \left(\sum_{j=1}^{n} N_{j,t}(h)\right)^2 \le n\sum_{j=1}^{n} N_{j,t}(h)^2 = nN_t(h).$$

Theorem 7.2 *For any $\varepsilon > 0$ and sufficiently large p for any non-negative h_1, h_2 with $h_1h_2 \le p$, the bound*

$$N_t(h_1h_2) \ge N_t(h_1) N_t(h_2)p^{-\varepsilon}$$

holds.

Proof We define the function

$$\psi(k) = 4\max_{m\le k} \tau^2(m).$$

Then $\psi(k) \leq k^\varepsilon$ for sufficiently large k.

We assume that $h_1 > 0$, $h_2 > 0$ because otherwise the inequality is trivial. Any solution (u_1, x_1, y_1) of the congruence

$$u_1 x_1 \equiv y_1 \pmod{p} \qquad 0 < |x_1|, |y_1| \leq h_1, \ u_1 \in V$$

and any solution (u_2, x_2, y_2) of the congruence

$$u_2 x_2 \equiv y_2 \pmod{p} \qquad 0 < |x_2|, |y_2| \leq h_2, \ u_2 \in V$$

defines the solution $(u = u_1 u_2, x = x_1 x_2, y = y_1 y_2)$ of the congruence

$$ux \equiv y \pmod{p} \qquad 0 < |x|, |y| \leq h_1 h_2, \ u \in V.$$

Also, because $h_1 h_2 \leq p$, any solution of the last equation corresponds to at most $\psi(h_1 h_2) \leq \psi(p)$ pairs of solutions of the two previous congruences. This implies the assertion of the theorem. $\qquad\square$

Corollary 7.3. *For any* h_1 *and* h_2 *such that* $1 \leq h_1 \leq h_2 < p$, *the following inequality holds:*

$$N_t(h_1) \ll \frac{h_1 N_t(h_2)}{h_2} p^\varepsilon.$$

Proof Denote $h_3 = \lfloor h_2 / h_1 \rfloor$. From (7.5) and Theorem 7.2 we obtain

$$\begin{aligned}
N_t(h_2) &\geq \ N_t(h_1 h_3) \gg N_t(h_1) N_t(h_3) p^{-\varepsilon} \\
&\geq \ N_t(h_1) h_3 p^{-\varepsilon} \gg \frac{N_t(h_1) h_2}{h_1} p^{-\varepsilon},
\end{aligned}$$

as required. $\qquad\square$

Theorem 7.4 *For any* $h \geq 1$ *the following inequality holds:*

$$N_t(h) \leq 225 \frac{h^2}{p} (t + N_t(\lfloor p/2h \rfloor)).$$

Proof It is easy to prove the theorem for $h > p/2$. Let us consider that $h < p/2$. We follow the ideas of the proof of Theorem 1.2 from [31] and Lemma 4 from [57]. We set

$$m = \left\lfloor 0.5 \left(\frac{p}{2h} - 1 \right) \right\rfloor, \qquad b_k = \frac{\pi^2 (2m + 1 - k)}{4(2m+1)^2}, \qquad |k| \leq 2m,$$

and define the functions

$$f(x) = \sum_{|k| \leq 2m} b_k \mathbf{e}(kx/p), \qquad F(x) = \sum_{u \in V} f(ux).$$

We have

$$0 \le b_k \le b_0 < 15h/p, \qquad |k| \le 2m. \qquad (7.6)$$

For any x we have,

$$f(x) = \frac{\pi^2}{4(2m+1)^2} \left| \sum_{s=0}^{2m} \mathbf{e}(sx/p) \right|^2 \ge 0.$$

Also we have

$$
\begin{aligned}
f(x) &= \frac{\pi^2}{4(2m+1)^2} \left| \sum_{s=0}^{2m} \mathbf{e}(sx/p) \right|^2 \\
&= \frac{\pi^2}{4(2m+1)^2} \left| \frac{\mathbf{e}\left((2m+1)x/p\right) - 1}{\mathbf{e}(x/p) - 1} \right|^2 \\
&= \left(\frac{\pi \sin\left(\pi(2m+1)x/p\right)}{2(2m+1)\sin(\pi x/p)} \right)^2.
\end{aligned}
$$

Using the inequalities

$$\frac{2}{\pi}|\psi| \le |\sin\psi| \le |\psi|$$

which holds for $-\pi/2 \le \psi \le \pi/2$, we deduce that $f(x) \ge 1$ for $|x| \le h \le p/2(2m+1)$.

From the previous inequalities, one can deduce

$$\sum_{x=0}^{p-1} F^2(x) \ge t N_t(h). \qquad (7.7)$$

As in [31], we denote by $w(k,l)$ the number of solutions of the congruence

$$k u_1 \equiv l u_2 \pmod{p}, \qquad u_1, u_2 \in V.$$

Taking into account (7.6), we get

$$
\begin{aligned}
\frac{1}{p} \sum_{x=0}^{p-1} F^2(x) &= \sum_{|k|,|l| \le 2m} b_k b_l w(k,l) \le \left(\frac{15h}{p} \right)^2 \sum_{|k|,|l| \le 2m} w(k,l) \\
&= \left(\frac{15h}{p} \right)^2 \left(w(0,0) + \sum_{1 \le |k|,|l| \le 2m} w(k,l) \right) \\
&\le \left(\frac{15h}{p} \right)^2 t(t + N_t(2m)).
\end{aligned}
$$

The combination of the last inequality with (7.7) completes the proof of the theorem. $\qquad \square$

Corollary 7.5. *If* $0 \le h \le p$ *and*

$$N_t(h) > 450\frac{h^2 t}{p},$$

then

$$N_t\left(\left\lfloor \frac{h^2}{p} \right\rfloor\right) < 450\frac{h^2}{p^{1-\varepsilon}}.$$

Proof It follows from Theorem 7.4 that

$$225\frac{h^2}{p}N_t(\lfloor p/2h \rfloor) \ge N_t(h) - 225\frac{h^2 t}{p} > N_t(h)/2,$$

or

$$N_t(h) < 450h^2 p^{-1} N_t(\lfloor p/2h \rfloor).$$

On the other hand, by Theorem 7.2,

$$N_t(h) \ge N_t\left(\left\lfloor \frac{p}{2h} \right\rfloor\right) N_t\left(\left\lfloor \frac{h^2}{p} \right\rfloor\right) p^{-\varepsilon}.$$

The two previous inequalities immediately imply the desired result.   □

Corollary 7.6. *For*

$$0 \le h \le \frac{p}{(450t)^2}$$

the bound

$$N_t(h) \ll hp^{\varepsilon}$$

holds.

Proof The assertion is evident for $h = 0$. Let $h > 0$ and $h_1 = 1 + \lfloor (hp)^{1/2} \rfloor$. Then

$$h_1 < 2(hp)^{1/2} \le \frac{p}{225t},$$

and, by (7.5),

$$N_t(h_1) \ge 2h_1 > 450\frac{h_1^2 t}{p}.$$

We can apply Corollary 7.5 to h_1. Setting $v = h_1^2/p$, we have

$$N_t(\lfloor v \rfloor) \ll vp^{\varepsilon},$$

and, therefore, since $h \leq \lfloor v \rfloor$, from Corollary 7.3, we obtain

$$N_t(h) \ll \frac{h N_t(\lfloor v \rfloor)}{\lfloor v \rfloor} \ll h p^{\varepsilon},$$

and we have the desired estimate.   □

Corollary 7.7. *For $t < p^{1/3}$ and $1 \leq h < p$, the bound*

$$N_t(h) \ll \max \left\{ h, \frac{h^2 t}{p} \right\} p^{\varepsilon}$$

holds.

Proof Let $h_0 = 450^2 \lceil p/t \rceil$. If $h \geq h_0$, then taking into account that $t < p^{1/3}$, we obtain

$$\lfloor p/2h \rfloor < t/450^2 \leq \frac{p}{(450t)^2}.$$

Now from Theorem 7.4 and Corollary 7.6,

$$N_t(h) \leq 225 h^2 p^{-1}(t + N_t(\lfloor p/2h \rfloor)) \ll h^2 p^{-1+\varepsilon}(t + p/h) \ll h^2 t p^{-1+\varepsilon}.$$

In particular, the last inequality holds for $h = h_0$, and we can also write

$$N_t(h_0) \ll h_0 p^{\varepsilon},$$

and, by Corollary 7.3, we get $N_t(h) \ll h p^{\varepsilon}$ for any $h \leq h_0$.   □

We believe that in fact the bound of Corollary 7.7 holds for any t.

Question 7.8. *Prove that for any t and $1 \leq h < p$, the bound*

$$N_t(h) \ll \max \left\{ h, \frac{h^2 t}{p} \right\} p^{\varepsilon}$$

holds.

We remark that Corollary 7.5 shows that for any $p^{1/2} \leq H < p$, this is true for at least one of the numbers $h = H$ or $h = \lfloor H^2/p \rfloor$. It is also true for $t < p^{1/3}$. Below we show that it holds for $t \geq p^{1/3}$ as well, provided that h is large enough.

Corollary 7.9. *For $t \geq p^{1/3}$ and $h \geq p^{3/4} t^{-1/4}$,*

$$N_t(h) \ll h^2 t p^{-1+\varepsilon}.$$

Proof Let us define

$$h_1 = \left\lfloor p^{1/2} t^{1/2} \right\rfloor, \qquad h_2 = \left\lfloor h_1^{1/2} \right\rfloor, \qquad h_3 = \lfloor p/2h \rfloor .$$

From Theorem 7.4, we obtain

$$N_t(h_1) \le 225 h_1^2 p^{-1} (t + N_t (\lfloor p/2h_1 \rfloor)).$$

Remarking that $\lfloor p/2h_1 \rfloor^2 \le h_1 \le p$ and from Theorem 7.2, we obtain

$$N_t (\lfloor p/2h_1 \rfloor) \le N_t \left(\lfloor p/2h_1 \rfloor^2 \right)^{1/2} p^\varepsilon \le N_t(h_1)^{1/2} p^\varepsilon ,$$

therefore

$$N_t(h_1) \ll h_1^2 p^{-1} \left(t + N_t(h_1)^{1/2} \right) p^{\varepsilon/2} \le t \left(t + N_t(h_1)^{1/2} \right) p^{\varepsilon/2} .$$

Hence $N_t(h_1) \le t^2 p^\varepsilon$.

Taking into account that $h_2^2 \le h_1$, by Theorem 7.2, we obtain

$$N_t(h_2) \ll N_t(h_1)^{1/2} p^{\varepsilon/2} \ll t p^\varepsilon .$$

Because $h_3 \le h_2$ for $h \ge p^{3/4} t^{-1/4}$, from Theorem 7.4, we see that

$$N_t(h) \le 225 \frac{h^2}{p} (t + N_t(h_3)) \le 225 \frac{h^2}{p} (t + N_t(h_2)).$$

From the previous inequality, we obtain the required result.   □

The above bounds can be combined and iterated in many possible ways to get the best possible estimate of $N_t(h)$ which then can be substituted in (7.2). Here we give only one example which is important for applications to the predictability of the $1/M$-pseudo-random number generator in Chapter 11.

Theorem 7.10 *For any* $\varepsilon > 0$ *and* $t \ge p^{1/2}$, *the bound*

$$H_p(t) \ll p^{34/37+\varepsilon}$$

holds.

Proof First of all we remark that for $t \ge 0.7 p^{2/3}$, the statement of the theorem follows from Lemma 7.1 and the bounds (7.3) and (7.5) applied with $h = \lfloor p^{3/37} \rfloor$ (because in this case $p^{1/2} N_t(h)^{1/2} \ll p^{1/2} h \ll p^{1/2+3/37} = o(p^{2/3})$).

Now assume that $t < 0.7 p^{2/3}$, thus we can use the bound (7.4). We also assume that ε is small enough.

For a sufficiently small $\delta > 0$, we define

$$L = \left\lceil 0.5 p^{20/37-\delta} \right\rceil, \qquad h = \left\lfloor L^2/p \right\rfloor$$

and remark that $h^6 < L$.

If

$$N_t(L) > 450\frac{L^2 t}{p},$$

then from Corollary 7.5 we obtain

$$N_t(h) \leq hp^\delta.$$

Otherwise, from Theorem 7.2 and Corollary 7.3 we derive

$$N_t(h)^6 \ll N_t(h^6)p^{\delta/2} \ll \frac{N_t(L)h^6}{L}p^\delta.$$

Therefore

$$N_t(h) \ll h(N_t(L)/L)^{1/6}p^{\delta/6} \ll h(Lt/p)^{1/6}p^{\delta/6} \ll ht^{1/6}p^{-17/222}.$$

Because $t^{1/6}p^{-17/222} \geq p^\delta$ for $t \geq p^{1/2}$ and sufficiently small δ, we may apply this bound in the first case as well. Thus we always have

$$N_t(h) \ll ht^{1/6}p^{-17/222} \ll p^{3/37}t^{1/6}p^{-17/222} = t^{1/6}p^{1/222}.$$

Substituting it in (7.4), we see that for any $k = 1, \dots, n$,

$$\sum_{j=1}^{n} N_{j,t}(h)|S_{j+k}(t)| \ll p^{1/4}h^{1/2}N_t(h)^{1/4}t^{3/8}$$

$$\ll p^{1/4}p^{3/74-\delta}t^{1/24}p^{1/888}t^{3/8}$$

$$= p^{7/24-\delta}t^{5/12} < 0.5t$$

for sufficiently large t. Now from Lemma 7.1, we derive that $H_p(t) \ll p^{1+\delta}h^{-1} \ll p^{34/37+3\delta}$. Putting $\delta = \varepsilon/3$, we obtain the desired results. $\qquad \square$

Unfortunately, we do not know if our result is the best possible even if t is of order $p^{1/2}$, even less do we know about other values of t. Moreover, we know examples showing that for some larger values of t, better estimates are possible. Finally, one can try to combine these considerations with estimates which follow from the Burgess bound of character sums (see Chapter 6 of [52]).

Question 7.11. *For each t, find the optimal values of α and the optimal way of combining the above estimates of $N_t(h)$ in order to get the smallest possible value on the right-hand side of (7.2).*

8

Multiplicative Translations of Arbitrary Sets Modulo p

In [31] the following problem has been considered, which is related to constructing and improving pseudo-random numbers.

For a set $R \subseteq \mathbb{F}_p$ and $a \in \mathbb{F}_p^*$, we denote by aR its multiplicative translation by a,

$$aR = \{au \; : \; u \in R\}$$

where multiplication means the multiplication over $\mathbb{F}_p$.

Then, for a set $R \subseteq \mathbb{F}_p$, we denote by $\mu(R)$ its smallest element

$$\mu(R) = \min_{u \in R} u$$

(as usual it is supposed that $\mathbb{F}_p = \{0, 1, \ldots, p-1\}$).

Finally, for a set $R \subseteq \mathbb{F}_p$, we define

$$M(R) = \frac{1}{p(p-1)} \sum_{a=1}^{p-1} \mu(aR).$$

We would like to know what can be said about this function $M(R)$ for

- arbitrary sets R;
- certain interesting sets R;
- specially constructed extremal sets.

First of all, we note that paper [31] provides the bounds

$$\frac{1}{2r} - \frac{1}{pr} \le M(R) \le \frac{100}{r^{1/2}}$$

which hold for any prime p and any set $R \subseteq \mathbb{F}_p$ of cardinality $|R| = r$, while 'on average' over all r-element sets $R \subseteq \mathbb{F}_p$,

$$\binom{p}{r}^{-1} \sum_{\substack{R \subseteq \mathbb{F}_p \\ |R| = r}} M(R) = \frac{1}{r+1}.$$

Also, for the special sets $I_r = \{1, \ldots, r\}$ and $J_r = \{\pm 1, \ldots, \pm t\}$ (here $r = 2t$),

$$M(I_r) = \alpha_r \frac{\ln r}{r} + O(rp^{-1}), \qquad M(J_r) = \beta_r \frac{1}{r} + O(rp^{-1})$$

where the constants α_r and β_r satisfy

$$\lim_{r \to \infty} \alpha_r = \frac{\pi^2}{24} = 0.41 \ldots, \qquad \lim_{r \to \infty} \beta_r = \frac{12 \ln 2}{\pi^2} = 0.84 \ldots.$$

Another interesting set considered in [31] is the set Q_r of quadratic non-residues modulo p where $r = (p-1)/2$. It was shown that $M(Q_r) = (1 + N_p)/2p$ where N_p is the smallest quadratic non-residue. Now one can apply all known results about the size of N_p. In particular, it was mentioned in [31] that known estimates of N_p imply that

$$M(Q_r) \gg \frac{\ln r \operatorname{Ln} \operatorname{Ln} \ln r}{r}$$

infinitely often. A conjecture was posed that $M(R) = O(r^{-1+\varepsilon})$ for any p and any r-element set $R \subseteq \mathbb{F}_p$. This conjecture is very strong, in particular, it implies the estimate $N_p = O(p^\varepsilon)$.

Here we show that for any p, one can find an r-element set R (with some $r > p/2$) satisfying a stronger estimate (note that the bigger the cardinality of R is the smaller $M(R)$ should be).

Theorem 8.1 *For any sufficiently large p, there exists $R \subseteq \mathbb{F}_p$ of cardinality $|R| = r > p/2$ and such that*

$$M(R) \geq 0.05 \frac{\ln r \operatorname{Ln} \ln r}{r}.$$

Proof Let $p_1, \ldots, p_s$ be all primes less than $n = \lfloor 0.4 \ln p \operatorname{Ln} \ln p \rfloor$, and let ϑ be a primitive root modulo p. Denote by $d_1, \ldots, d_s$ the discrete logarithms of $p_1, \ldots, p_s$ in base ϑ, that is, $\vartheta^{d_j} \equiv p_j \pmod{p}$, $j = 1, \ldots, s$. As before, for real u, we denote by $\|u\|$ the distance from u to the nearest integer. We have

$$\prod_{j=1}^{s} \frac{10\ln n}{\ln p_j} \le (10\ln n)^s \prod_{p_j > n/\ln^2 n} \frac{1}{\ln p_j}$$

$$\le (10\ln n)^s (\ln n - 2\operatorname{Ln}\ln n)^{-s+n/\ln^2 n}$$

$$\le 10^s (1 - 2\operatorname{Ln}\ln n/\ln n)^{-s} (\ln n)^{n/\ln^2 n}$$

$$= 10^{s+o(s)} = 10^{0.4\ln p + o(\ln p)} = p^{0.4\ln 10 + o(1)} < p^{0.93},$$

provided that p is large enough.

Therefore, by Dirichlet's principle, there exists a positive integer $k < p^{0.93}$ such that

$$\left\|\frac{d_j k}{p-1}\right\| < \frac{\ln p_j}{10\ln n}, \quad j = 1, \dots, s.$$

Assume that $\vartheta^d \equiv m \pmod{p}$ with $1 \le m \le n$ and let α_j be the largest power of p_j dividing m, $j = 1, \dots, s$. We have

$$\left\|\frac{dk}{p-1}\right\| \le \sum_{j=1}^{s} \alpha_j \left\|\frac{d_j k}{p-1}\right\| < \sum_{j=1}^{s} \alpha_j \frac{\ln p_j}{10\ln n} = \frac{\ln m}{10\ln n} \le 0.1.$$

Denote

$$R = \left\{\vartheta^d : \left\|\frac{dk}{p-1}\right\| > 0.24\right\}, \qquad A = \left\{\vartheta^d : \left\|\frac{dk}{p-1}\right\| < 0.14\right\}.$$

We have $r = |R| = 0.52p + o(p) > 0.5p$ and $|A| = 0.28p + o(p) > 0.25p$ provided that p is large enough. Because of the choice of k, we obtain $\mu(aR) > n$ for $a \in A$. Therefore,

$$M(R) > \frac{|A|(n+1)}{p(p-1)} \ge \frac{0.25(n+1)}{p} \ge \frac{n+1}{8r}$$

and the estimate follows. □

We finish this chapter with a remark that the bit complexity of the construction of the set R of Theorem 8.1 is $O(p\ln^{1+\varepsilon} p)$, that is $O(\ln^{1+\varepsilon} p)$ per element.

Part four

Applications to Algebraic Number Fields

9

Representatives of Residue Classes

Let $\mathfrak{q}$ be an integer ideal of an algebraic number field $\mathbb{K}$. For a residue class $\alpha \in \Lambda_{\mathfrak{q}}$, denote by $N_{\mathfrak{q}}(\alpha)$ the minimal norm of all elements of α,

$$N_{\mathfrak{q}}(\alpha) = \min_{a \in \alpha} | \operatorname{Nm}(a)|,$$

and let $L(\mathbb{K}, \mathfrak{q})$ and $A(\mathbb{K}, \mathfrak{q})$ be the largest and average values of $N_{\mathfrak{q}}(\alpha)$ over all residue classes of $\Lambda_{\mathfrak{q}}^*$. That is, we set

$$L(\mathbb{K}, \mathfrak{q}) = \max_{\alpha \in \Lambda_{\mathfrak{q}}^*} N_{\mathfrak{q}}(\alpha), \qquad A(\mathbb{K}, \mathfrak{q}) = \frac{1}{\varphi(\mathfrak{q})} \sum_{\alpha \in \Lambda_{\mathfrak{q}}^*} N_{\mathfrak{q}}(\alpha).$$

It is easy to see that $A(\mathbb{K}, \mathfrak{q}) \leq L(\mathbb{K}, \mathfrak{q}) = O(\operatorname{Nm}(\mathfrak{q}))$. Upper bounds for the implied constant in this estimate are obtained in [11] where some connections of this problem with the theory of diophantine approximation are displayed.

We also note that the inequality

$$L(\mathbb{K}, \mathfrak{q}) < \operatorname{Nm}(\mathfrak{q})$$

(even for principal ideals only) would mean that $\mathbb{K}$ is Euclidean with respect to its norm (so far very few examples of such fields are known, see [49, 50, 65, 70], a complete catalogue of known Euclidean number fields is given in [48]). On the other hand, upper bounds on $L(\mathbb{K}, \mathfrak{q})$ and $A(\mathbb{K}, \mathfrak{q})$ show that perhaps, in some sense, any algebraic number field is 'nearly' Euclidean or is at least Euclidean 'on average'.

It is interesting to note that a criterion for Euclidicity (with respect to the field norm) also relies on a certain property of finitely generated groups (see [48, 49, 65, 70] and Chapter 16 of this book). Moreover, Euclidicity with respect to some other function (a certain modification of the norm) is related to Artin's conjecture and its modifications (see [50]).

Now suppose that $\mathbb{K}$ has $r \geq 1$ principal units; that is, that $\mathbb{K}$ is neither the field of rationals $\mathbb{Q}$ nor an imaginary quadratic extension of $\mathbb{Q}$.

65

It is shown in [12] that $A(\mathbb{K}, p) = o(p^n)$ for almost all p in the sense of the asymptotic density of rational prime numbers p. From here on, we identify integer algebraic numbers and the corresponding principal ideals. Thus,

$$A(\mathbb{K}, \mathfrak{q}) = o(\mathrm{Nm}(\mathfrak{q})) \tag{9.1}$$

for some infinite sequence of integer ideals $\mathfrak{q}$. In [12] this result is formulated in a slightly different but equivalent form.

Thus in fields with $r \geq 1$, the behavior of $A(\mathbb{K}, \mathfrak{q})$ differs from that of $A(\mathbb{Q}, q)$ with integers $q \geq 2$, for there, one has

$$A(\mathbb{Q}, q) = q^{-1} \sum_{-q/2 \leq a < q/2} |a| = q/4 + O(1).$$

The results of [12] are improved in [79]. It is shown in that paper that $A(\mathbb{K}, p) = O(p^n (\ln p)^{-1/3})$ and that $A(\mathbb{K}, p) = O(p^{n-1/6}(\ln p)^{1/3})$ for all and almost all rational prime numbers p, respectively. Thus in place of (9.1), we see that there is an infinite sequence of integer ideals $\mathfrak{q}$ with

$$A(\mathbb{K}, \mathfrak{q}) = O\left(\mathrm{Nm}(\mathfrak{q})^{1-1/6n} (\ln \mathrm{Nm}(\mathfrak{q}))^{1/3}\right). \tag{9.2}$$

Here we consider prime ideals $\mathfrak{p}$ and their powers $\mathfrak{p}^k$ instead of just principal ideals (p). This allows us to get bounds which are much stronger than (9.1) and (9.2). Moreover, for almost all prime ideals $\mathfrak{p}$, we obtain a non-trivial upper bound for $L(\mathbb{K}, \mathfrak{p})$ as well.

Although we shall apply Lemmas 9.1 and 9.2 to prime ideals of the first degree, these results are stated for an arbitrary prime ideal.

Throughout this chapter, U denotes the group of principal units of $\mathbb{Z}_\mathbb{K}$.

Lemma 9.1. *For any prime ideal $\mathfrak{p}$, the bound*

$$A(\mathbb{K}, \mathfrak{p}) = O(\mathrm{Nm}(\mathfrak{p})|U_\mathfrak{p}|^{-1/2})$$

holds.

Proof Fix an integral basis $\omega_1, \ldots, \omega_n$ of $\mathbb{Z}_\mathbb{K}$ over $\mathbb{Z}$ (see Theorem 2.5 of [65]), and for $h > 0$, denote by $\mathfrak{B}$ the box

$$\mathfrak{B} = \{z = z_1\omega_1 + \cdots + z_n\omega_n \in \mathbb{Z}_\mathbb{K} : |z_i| \leq h, \ i = 1, \ldots, n\}.$$

For an arbitrary fixed primitive additive character χ of $\Lambda_\mathfrak{p}$ and $\lambda \in \Lambda_\mathfrak{p}$, we define the sums

$$S(\lambda) = \sum_{u \in U_\mathfrak{p}} \chi(\lambda u), \quad T(\lambda, h) = \sum_{z \in \mathfrak{B}} \chi(\lambda z).$$

In the sequel we make repeated use of the well-known result

$$\sum_{\lambda \in \Lambda_{\mathfrak{p}}} \chi(\lambda \vartheta) = \begin{cases} \text{Nm}(\mathfrak{p}), & \text{if } \vartheta \equiv 0 \pmod{\mathfrak{p}}, \\ 0, & \text{otherwise.} \end{cases}$$

Denote by $T_\alpha(U, \mathfrak{p}, h)$ the number of solutions of the congruence

$$\alpha u \equiv x - y \pmod{\mathfrak{p}}, \qquad u \in U_{\mathfrak{p}}, \ x, y \in \mathfrak{B}, \tag{9.3}$$

and let $R(U, \mathfrak{p}, h)$ be the set of $\alpha \in \Lambda_{\mathfrak{p}}^*$ with $T_\alpha(U, \mathfrak{p}, h) = 0$. We shall now show that

$$|R(U, \mathfrak{p}, h)| = O\left(\text{Nm}(\mathfrak{p})|U_{\mathfrak{p}}|^{-1}(1 + \text{Nm}(\mathfrak{p})^2/h^{2n})\right). \tag{9.4}$$

Indeed, we have

$$\begin{aligned} T_\alpha(U, \mathfrak{p}, h) &= \text{Nm}(\mathfrak{p})^{-1} \sum_{u \in U_{\mathfrak{p}}} \sum_{x, y \in \mathfrak{B}} \sum_{\lambda \in \Lambda_{\mathfrak{p}}} \chi\left(\lambda(\alpha u - x + y)\right) \\ &= |\mathfrak{B}|^2 |U_{\mathfrak{p}}|/\text{Nm}(\mathfrak{p}) + \text{Nm}(\mathfrak{p})^{-1} \sum_{\lambda \in \Lambda_{\mathfrak{p}}^*} S(\alpha\lambda)\, |T(\lambda, h)|^2. \end{aligned}$$

Summing this equation over all $\alpha \in R(U, \mathfrak{p}, h)$ we get

$$|R(U, \mathfrak{p}, h)| \le W/|\mathfrak{B}|^2 |U_{\mathfrak{p}}|$$

where

$$\begin{aligned} W &= \sum_{\alpha \in R(U, \mathfrak{p}, h)} \sum_{\lambda \in \Lambda_{\mathfrak{p}}^*} |S(\alpha\lambda)||T(\lambda, h)|^2 \\ &= \sum_{\lambda \in \Lambda_{\mathfrak{p}}^*} |T(\lambda, h)|^2 \sum_{\alpha \in R(U, \mathfrak{p}, h)} |S(\alpha\lambda)|. \end{aligned}$$

Then, from (3.1) we get

$$\begin{aligned} \sum_{\alpha \in R(U, \mathfrak{p}, h)} |S(\alpha\lambda)| &\le \left(|R(U, \mathfrak{p}, h)| \sum_{\alpha \in \Lambda_{\mathfrak{p}}} |S(\alpha\lambda)|^2 \right)^{1/2} \\ &= \left(|R(U, \mathfrak{p}, h)| \sum_{\alpha \in \Lambda_{\mathfrak{p}}} |S(\alpha)|^2 \right)^{1/2} \\ &= (|R(U, \mathfrak{p}, h)| \, \text{Nm}(\mathfrak{p})|U_{\mathfrak{p}}|)^{1/2}. \end{aligned}$$

Therefore,

$$W \le (|R(U, \mathfrak{p}, h)| \, \text{Nm}(\mathfrak{p})|U_{\mathfrak{p}}|)^{1/2} \sum_{\lambda \in \Lambda_{\mathfrak{p}}} |T(\lambda, h)|^2.$$

It is easy to see that

$$\sum_{\lambda \in \Lambda_\mathfrak{p}} |T(\lambda, h)|^2 = \mathrm{Nm}(\mathfrak{p}) Q(h)$$

where $Q(h)$ is the number of solutions of the congruence

$$x \equiv y \pmod{\mathfrak{p}}, \qquad x, y \in \mathfrak{B}.$$

This can be estimated as $O(|\mathfrak{B}|(|\mathfrak{B}|/\mathrm{Nm}(\mathfrak{p}) + 1))$ hence

$$\sum_{\lambda \in \Lambda_\mathfrak{p}} |T(\lambda, h)|^2 = O\left(|\mathfrak{B}|^2 + |\mathfrak{B}|\, \mathrm{Nm}(\mathfrak{p})\right). \tag{9.5}$$

This implies

$$W = O\left(\left(|R(U, \mathfrak{p}, h)|\, \mathrm{Nm}(\mathfrak{p})|U_\mathfrak{p}|\right)^{1/2} \left(|\mathfrak{B}|^2 + |\mathfrak{B}|\, \mathrm{Nm}(\mathfrak{p})\right)\right)$$

and we get the estimate (9.4).

Now, let $l_\mathfrak{p}(N)$ and $L_\mathfrak{p}(N)$ be the number of classes $\alpha \in \Lambda_\mathfrak{p}$ with $N_\mathfrak{p}(\alpha) = N$ and $N_\mathfrak{p}(\alpha) > N$, respectively.

Denote by $H(\vartheta)$ the height of $\vartheta \in \mathbb{Z}_\mathbb{K}$, that is the largest absolute value of coordinates in the representation of ϑ with respect to the basis $\omega_1, \ldots, \omega_n$. We have

$$|\mathrm{Nm}(\vartheta)| \leq (n\omega H(\vartheta))^n,$$

where ω is the maximum of absolute values of $\omega_1, \ldots, \omega_n$ and of all their conjugates over $\mathbb{Q}$. Then, it follows from the property

$$N_\mathfrak{p}(\alpha) = N_\mathfrak{p}(\alpha u), \quad u \in U_\mathfrak{p},$$

that if the congruence (9.3) is solvable for some $\alpha \in \Lambda_\mathfrak{p}$, then $N_\mathfrak{p}(\alpha) \leq (n\omega h)^n$. Therefore

$$L_\mathfrak{p}(N) \leq |R(U, \mathfrak{p}, n^{-1}\omega^{-1}N^{1/n})|.$$

Using (9.4) we obtain

$$L_\mathfrak{p}(N) = O\left(\mathrm{Nm}(\mathfrak{p})|U_\mathfrak{p}|^{-1}(1 + \mathrm{Nm}(\mathfrak{p})^2/N^2)\right) \tag{9.6}$$

for all $N > 0$. Finally, we have

$$
\begin{aligned}
A(\mathbb{K}, \mathfrak{p}) &= \operatorname{Nm}(\mathfrak{p})^{-1} \sum_{N=1}^{\operatorname{Nm}(\mathfrak{p})-1} l_\mathfrak{p}(N) N \\
&= \operatorname{Nm}(\mathfrak{p})^{-1} \sum_{N=1}^{\operatorname{Nm}(\mathfrak{p})-1} \left(L_\mathfrak{p}(N-1) - L_\mathfrak{p}(N) \right) N \\
&\leq \operatorname{Nm}(\mathfrak{p})^{-1} \sum_{N=0}^{\operatorname{Nm}(\mathfrak{p})-1} L_\mathfrak{p}(N).
\end{aligned}
$$

Applying (9.6) if $N > \operatorname{Nm}(\mathfrak{p})|U_\mathfrak{p}|^{-1/2}$ and the trivial bound $L_\mathfrak{p}(N) \leq \operatorname{Nm}(\mathfrak{p})$ otherwise, we obtain the result claimed. $\qquad\square$

We note that using Lemma 3.3 instead of (3.1), for a prime ideal $\mathfrak{p}$ of first degree one can get the estimate

$$
|R(U, \mathfrak{p}, h)| = O\left(\operatorname{Nm}(\mathfrak{p})|U_\mathfrak{p}|^{-3/2}(1 + \operatorname{Nm}(\mathfrak{p})^4/h^{4n}) \right)
$$

which improves (9.4) for some values of parameters but unfortunately does not improve Lemma 9.1.

Lemma 9.2. *For any prime ideal $\mathfrak{p}$, the bound*

$$
L(\mathbb{K}, \mathfrak{p}) = O(\operatorname{Nm}(\mathfrak{p})^{3/2}|U_\mathfrak{p}|^{-1})
$$

holds.

Proof We use the same notation as in the proof of Lemma 9.1. Proceeding as before and applying (3.15) and (9.5), we get

$$
\begin{aligned}
\left| T_\alpha(U, \mathfrak{p}, h) - \frac{|\mathfrak{B}|^2|U_\mathfrak{p}|}{\operatorname{Nm}(\mathfrak{p})} \right| &\leq \operatorname{Nm}(\mathfrak{p})^{-1} \sum_{\lambda \in \Lambda_\mathfrak{p}^*} |S(\alpha\lambda)| |T(\lambda, h)|^2 \\
&\leq \operatorname{Nm}(\mathfrak{p})^{-1/2} \sum_{\lambda \in \Lambda_\mathfrak{p}^*} |T(\lambda, h)|^2 \\
&= O\left(\operatorname{Nm}(\mathfrak{p})^{-1/2}|\mathfrak{B}|^2 + \operatorname{Nm}(\mathfrak{p})^{1/2}|\mathfrak{B}| \right).
\end{aligned}
$$

We may suppose that $|U_\mathfrak{p}| > \operatorname{Nm}(\mathfrak{p})^{1/2}$, since otherwise the bound is trivial. Then there is a constant $c > 0$ such that $T_\alpha(U, \mathfrak{p}, h) > 0$ for $h = c(\operatorname{Nm}(\mathfrak{p})^{3/2}|U_\mathfrak{p}|^{-1})^{1/n}$. Thus, $N_\mathfrak{p}(\alpha) = O(h^n) = O(\operatorname{Nm}(\mathfrak{p})^{3/2}|U_\mathfrak{p}|^{-1})$ for any $\alpha \in \Lambda_\mathfrak{p}$. $\qquad\square$

Lemma 9.3. *For any prime ideal $\mathfrak{p}$ of first degree, the bound*

$$L(\mathbb{K}, \mathfrak{p}) = O\left(\min\{\mathrm{Nm}(\mathfrak{p})^{5/4}|U_\mathfrak{p}|^{-5/8}, \; \mathrm{Nm}(\mathfrak{p})^{9/8}|U_\mathfrak{p}|^{-3/8}\}\right)$$

holds.

Proof Proceeding as in the proof of Lemma 9.2 and applying the bounds (3.16) and (3.17) instead of (3.15) in the corresponding place, we get the desired result. $\qquad\square$

Lemma 9.4. *For any unramified fixed prime ideal $\mathfrak{p}$ of first degree, the bound*

$$L(\mathbb{K}, \mathfrak{p}^k) = O(\mathrm{Nm}(\mathfrak{p}^k)|U_{\mathfrak{p}^k}|^{-1})$$

holds.

Proof Proceeding as in the proof of Lemma 9.2 and applying the bound (3.22) instead of (3.15) in the corresponding place we get the desired result. $\qquad\square$

Lemma 9.5. *For any group V and any integer ideal $\mathfrak{q}$ which is relatively prime to each generator $\lambda_1, \ldots, \lambda_r$, the bound*

$$|V_\mathfrak{q}| \gg (\ln \mathrm{Nm}(\mathfrak{q}))^r$$

holds.

Proof Let L be the maximum of the absolute values of $\lambda_1, \ldots, \lambda_r$ and of all their conjugates over $\mathbb{Q}$. Evidently, for

$$Q = \left\lfloor \frac{\ln \mathrm{Nm}(\mathfrak{q})}{rn \ln(2L+1)} \right\rfloor$$

all $Q^r \gg (\ln \mathrm{Nm}(\mathfrak{q}))^r$ numbers

$$\{\lambda_1^{x_1} \ldots \lambda_r^{x_r} \; : \; 0 \le x_1, \ldots, x_r \le Q-1\}$$

are distinct modulo $\mathfrak{q}$ because they are distinct elements of $\mathbb{Z}_\mathbb{K}$ and norms of their differences do not exceed $(2L^rQ)^n < \mathrm{Nm}(\mathfrak{q})$. $\qquad\square$

We also need the following well-known statement that is Corollary 2 of Proposition 7.10 of [65] (showing that 'almost all' prime ideals are of the first degree).

Lemma 9.6. *The number of prime ideals $\mathfrak{p}$ of first degree such that $\mathrm{Nm}(\mathfrak{p}) \le N$ equals $N/\ln N + o(N/\ln N)$.*

In the next lemma we show that, for almost all prime ideals $\mathfrak{p}$, $|U_\mathfrak{p}|$ can be much better estimated. Actually, this is a generalization of Theorem 2.1. of [72] and is obtained by using the same ideas.

Lemma 9.7. *For any $\rho < (1 - \ln 2)/2$ and any group V, the number of prime ideals $\mathfrak{p}$ with $\mathrm{Nm}(\mathfrak{p}) \le N$ for which*

$$|V_\mathfrak{p}| < \mathrm{Nm}(\mathfrak{p})^{r/(r+1)} \exp\left(\ln^\rho \mathrm{Nm}(\mathfrak{p})\right)$$

is $o(N/\ln N)$.

Proof As we see from Lemma 9.6, it is enough to consider only prime ideals $\mathfrak{p}$ of first degree.

For some integer $T > 0$, consider the product

$$Q(T) = \prod_{0 \le x_1,\dots,x_r \le T} \prod_{I,J} \left(\prod_{i \in I} \lambda_i^{x_i} - \prod_{j \in J} \lambda_j^{x_j} \right)$$

where I, J run over all 2^r disjoint partitions of the set of indices $\mathbb{N}_r = \{1, \dots, r\}$; that is $I \cup J = \mathbb{N}_r$ and $I \cap J = \emptyset$.

It is evident that if $|U_\mathfrak{p}| < (T+1)^r$, then among the $(T+1)^r$ elements

$$\{\epsilon_1^{x_1} \dots \epsilon_r^{x_r} : 0 \le x_1, \dots, x_r \le T\}$$

there are at least two elements in the same residue class modulo $\mathfrak{p}$. Hence $\mathfrak{p} \mid Q(T)$ and $\mathrm{Nm}(\mathfrak{p}) \mid \mathrm{Nm}(Q(T))$.

On the other hand, we have $\mathrm{Nm}(Q(T)) = \exp\left(O(T^{r+1})\right)$. Therefore $\mathrm{Nm}(Q(T))$ has at most $O(T^{r+1}/\ln T)$ rational prime divisors.
Set

$$T = \left\lfloor \left(\frac{N}{\psi(N)} \right)^{1/(r+1)} \right\rfloor$$

where $\psi(N) \to \infty$ is some monotonically increasing function. Then we see that there are at most $O(N/\psi(N) \ln N)$ rational prime numbers $p \le N$ having a prime ideal divisor $\mathfrak{p}$ with

$$|U_\mathfrak{p}| < \left(\frac{p}{\psi(p)} \right)^{r/(r+1)} \le (T+1)^r.$$

Also, it is shown in the proof of Theorem 2.1 of [72] that there is a function $\psi(N) \to \infty$ such that the number of primes $p \le N$ such that $p - 1$ has an integer divisor d in the range

$$\left(\frac{p}{\psi(p) \ln p} \right)^{r/(r+1)} \le d \le p^{r/(r+1)} \exp(\ln^\rho p)$$

is $o(\pi(N))$ for any $\rho < (1 - \ln 2)/2$.

Taking into account that $|U_\mathfrak{p}| \mid p - 1$ (since $\mathfrak{p}$ is of first degree), we obtain the statement.   □

Lemma 9.8. *For any group V and unramified fixed prime ideal $\mathfrak{p}$ of first degree which is relatively prime to $\lambda_1, \dots, \lambda_r$, there is a constant $C(\mathfrak{p}) > 0$ such that the bound*

$$|V_{\mathfrak{p}^k}| \geq C(\mathfrak{p}) \, \mathrm{Nm}(\mathfrak{p}^k)$$

holds.

Proof If $r = 1$, this result is known in the following form. If t_k is the multiplicative order of λ_1 modulo $\mathfrak{p}^k$ and $t_1 = \cdots = t_\alpha \neq t_{\alpha+1}$, then $t_k = t_1 \, \mathrm{Nm}(\mathfrak{p}^{k-\alpha})$. Clearly, adding several more generators may only extend the group.   □

Now, Lemmas 9.1 and 9.5 yield:

Theorem 9.9 *Let $\mathbb{K}$ have $r \geq 1$ multiplicatively independent principal units and let $\mathfrak{p}$ be a prime ideal. Then*

$$A(\mathbb{K}, \mathfrak{p}) = O\left(\mathrm{Nm}(\mathfrak{p}) \, (\ln \mathrm{Nm}(\mathfrak{p}))^{-r/2}\right).$$

Lemmas 9.1, 9.2, 9.3, and 9.7 yield:

Theorem 9.10 *Let $\mathbb{K}$ have $r \geq 1$ multiplicatively independent principal units. Then for any fixed $\rho < (1 - \ln 2)/2$, the bounds*

$$L(\mathbb{K}, \mathfrak{p}) = \begin{cases} O\left(\mathrm{Nm}(\mathfrak{p})^{1/2+1/(r+1)} \exp\left(-\ln^\rho \mathrm{Nm}(\mathfrak{p})\right)\right), & \text{if } r \geq 2, \\ O\left(\mathrm{Nm}(\mathfrak{p})^{15/16} \exp\left(-\tfrac{5}{8} \ln^\rho \mathrm{Nm}(\mathfrak{p})\right)\right), & \text{if } r = 1, \end{cases}$$

and

$$A(\mathbb{K}, \mathfrak{p}) = O\left(\mathrm{Nm}(\mathfrak{p})^{1/2+1/2(r+1)} \exp\left(-0.5 \ln^\rho \mathrm{Nm}(\mathfrak{p})\right)\right)$$

hold for a sequence of prime ideals $\mathfrak{p}$ of asymptotic density 1.

It is evident that $r + 1 \geq n/2$; see Theorem 3.6 of [65]. Thus the estimates of Theorem 9.10 can be rewritten in the form

$$L(\mathbb{K}, \mathfrak{p}) = o\left(\mathrm{Nm}(\mathfrak{p})^{1/2+2/n}\right), \qquad A(\mathbb{K}, \mathfrak{p}) = o\left(\mathrm{Nm}(\mathfrak{p})^{1/2+1/n}\right).$$

Finally, Lemmas 9.4 and 9.8 yield:

Theorem 9.11 *Suppose that* $\mathbb{K}$ *has* $r \geq 1$ *multiplicatively independent principal units and that* $\mathfrak{p}$ *is an unramified fixed prime ideal of first degree. Then* $L(\mathbb{K}, \mathfrak{p}^k) = O(1)$.

In fact, the bound (3.23) infers that $L(\mathbb{K}, \mathfrak{p}^k) = O(\mathrm{Nm}(\mathfrak{p}^k)^{1-1/d+\varepsilon})$ for any unramified fixed prime ideal $\mathfrak{p}$ of degree d.

Now let us make a few comments.

First of all, let us stress links of problems considered here with an appropriate version of Artin's conjecture for algebraic number fields. Indeed, if $U_{\mathfrak{p}} = \Lambda_{\mathfrak{p}}^*$ for an infinite sequence of prime ideals $\mathfrak{p}$ (this can be considered as a modification of Artin's conjecture) then $L(\mathbb{K}, \mathfrak{p}) = 1$ for this sequence.

Moreover, it follows from the results of [32] that in the case of $\mathbb{K} = \mathbb{Q}$ for any group V with $r \geq 3$ multiplicatively independent generators, $|V_p| = p-1$ for infinitely many rational prime numbers p. Unfortunately, for algebraic number fields (even for the particular case of the unit group U), no analogue of this result is known.

Note that quite simple elementary considerations allow us to get the lower bound

$$L(\mathbb{K}, \mathfrak{q}) \geq A(\mathbb{K}, \mathfrak{q}) \gg \mathrm{Nm}(\mathfrak{q})/|U_{\mathfrak{q}}| \, (\mathrm{Ln}\ln\mathrm{Nm}(\mathfrak{q}))^{n-1} \,, \qquad (9.7)$$

showing (together with Lemma 9.1) that the behavior of $L(\mathbb{K}, \mathfrak{q})$ and $A(\mathbb{K}, \mathfrak{q})$ does depend on the size of $U_{\mathfrak{q}}$. To show this, denote by $d_n(N)$ the number of representations of integer $N > 0$ as a product of n different positive integers. It follows from Lemma 4.2 of [65] that there exist at most $d_n(N)$ integer ideals $\mathfrak{q}$ with $\mathrm{Nm}(\mathfrak{q}) = N$. Thus there exist at most $l_{\mathfrak{q}}(N) = O(|U_{\mathfrak{q}}|d_n(N))$ pairwise different residue classes $\alpha \in \Lambda_{\mathfrak{q}}$ with $N_{\mathfrak{q}}(\alpha) = N$. Using the two following well-known estimates

$$\sum_{N=1}^{Q} d_n(N) = O(Q(\mathrm{Ln}\ln Q)^{n-1}), \qquad \varphi(\mathfrak{q}) \gg \mathrm{Nm}(\mathfrak{q})/\mathrm{Ln}\ln\mathrm{Nm}(\mathfrak{q})$$

and the evident equality

$$\sum_{N=1}^{\mathrm{Nm}(\mathfrak{q})-1} l_{\mathfrak{q}}(N) = \varphi(\mathfrak{q}) - 1,$$

we get

$$A(\mathbb{K}, \mathfrak{q}) = \varphi(\mathfrak{q})^{-1} \sum_{N=1}^{\mathrm{Nm}(\mathfrak{q})-1} l_{\mathfrak{q}}(N)N$$

$$= \varphi(\mathfrak{q})^{-1} \sum_{Q=1}^{\mathrm{Nm}(\mathfrak{q})-1} \left(\varphi(\mathfrak{q}) - 1 - \sum_{N=1}^{Q-1} l_{\mathfrak{q}}(N) \right)$$

$$= \quad \varphi(\mathfrak{q})^{-1} \sum_{Q=1}^{\mathrm{Nm}(\mathfrak{q})-1} \min\{0, \ \varphi(\mathfrak{q}) - O(|U_{\mathfrak{q}}|Q(\mathrm{Ln} \ln Q)^{n-1})\}$$

$$\gg \quad \varphi(\mathfrak{q})/|U_{\mathfrak{q}}| (\mathrm{Ln} \ln \varphi(\mathfrak{q}))^{n-1}$$

$$\gg \quad \mathrm{Nm}(\mathfrak{q})/|U_{\mathfrak{q}}| (\mathrm{Ln} \ln \mathrm{Nm}(\mathfrak{q}))^{n} \ .$$

In particular, one can derive from this bound that, for $r = 1$ (that is, $n \leq 4$), $A(\mathbb{K}, \mathfrak{q})$ can be large enough. Indeed, let us consider the following sequence of principal ideals $\mathfrak{q}_k = (\epsilon^k - 1)$, $k = 1, 2, \ldots$, where ϵ is a principal unit of $\mathbb{K}$. It is easy to prove that $|U_{\mathfrak{q}_k}| = O(k) = O(\ln \mathrm{Nm}(\mathfrak{q}_k))$ for such ideals, hence

$$A(\mathbb{K}, \mathfrak{q}) \gg \mathrm{Nm}(\mathfrak{q})/\ln \mathrm{Nm}(\mathfrak{q}) (\mathrm{Ln} \ln \mathrm{Nm}(\mathfrak{q}))^{n}$$

for an infinite sequence of integer ideals.

We do not know any other non-trivial general lower bounds for $L(\mathbb{K}, \mathfrak{q})$ and $A(\mathbb{K}, \mathfrak{q})$ (in terms of $\mathrm{Nm}(\mathfrak{q})$ only). Any of them would be very interesting.

One sees that there is a large gap between (9.7) and the upper bound of Lemma 9.1, but (9.7) seems to be more precise.

Question 9.12. *Extend Lemma 9.1 to arbitrary integer ideal* $\mathfrak{q}$.

There are some difficulties with zero divisors of $\Lambda_{\mathfrak{q}}^*$ but none the less we believe it is possible to obtain the generalization. If Lemma 9.1 were generalized in this way, one could get an analogue of Lemma 9.7 for almost all integer ideals and the estimate of $A(\mathbb{K}, \mathfrak{q})$ would be better than that which would simply follow from Lemma 9.5.

Question 9.13. *Does the bound*

$$A(\mathbb{K}, \mathfrak{q}) = O\left(\mathrm{Nm}(\mathfrak{q})|U_{\mathfrak{q}}|^{-1+\delta}\right)$$

hold for an arbitrary integer ideal $\mathfrak{q}$ *and any* $\delta > 0$ *(at least for a prime ideal* $\mathfrak{p}$)?

As we have mentioned, one cannot hope to get even the following 'weak' estimate $L(\mathbb{K}, \mathfrak{q}) = o(\mathrm{Nm}(\mathfrak{q}))$ for all integer ideals. However, we do believe that the answer to the following question is positive.

Question 9.14. *Does the bound*

$$L(\mathbb{K}, \mathfrak{q}) = o(\mathrm{Nm}(\mathfrak{q}))$$

hold for 'almost all' integer ideals $\mathfrak{q}$?

The following question (actually three separate questions) seems to be really hard.

Question 9.15. *Find the distribution of* $A(\mathbb{K}, \mathfrak{q})$ *and* $L(\mathbb{K}, \mathfrak{q})$ *on the sets of*

- *all integer ideals;*
- *all prime ideals;*
- *all principal ideals; and on other interesting sets.*

If we could prove that $L(\mathbb{K}, \mathfrak{q}) = o(\mathrm{Nm}(\mathfrak{q}))$ (or at least $A(\mathbb{K}, \mathfrak{q}) = o(\mathrm{Nm}(\mathfrak{q}))$) for almost all principal ideals, it would mean that any algebraic number field is 'nearly' Euclidean and moreover the analogue of Euclid's algorithm would work and would even run in a sub-logarithmic number of steps for the majority of inputs. That makes the situation quite different from that of Euclid's algorithm for integers, where it performs a logarithmic number of steps in both worst and average cases.

Finally, we note that if we keep in mind applications to Euclid's algorithm, then the following question is of interest.

Question 9.16. *How fast can one find* $a \in \alpha$ *with* $|\mathrm{Nm}(a)| = N_{\mathfrak{q}}(\alpha)$ *for any (or 'almost any') integer ideal* $\mathfrak{q}$ *and any (or 'almost any') residue class* $\alpha \in \Lambda_{\mathfrak{q}}^{*}$?

A detailed exhibition of many problems related to the Euclid algorithm in algebraic number fields is given in [8].

10

Cyclotomic Fields and Gaussian Periods

Let p be a prime, and let $g \geq 2$ be a fixed primitive root modulo p.

In papers [20, 21, 22] relations were discovered between the distribution of the g-ary digits of $1/p$ and some properties of the minus part $h^-(p, s)$ of the class numbers of the subfield $\mathbb{K}_s \subseteq \mathbb{Q}(e(1/p))$ of degree $[\mathbb{K}_s : \mathbb{Q}] = s$ where s is an even divisor of $p - 1$ but $2s$ is not.

Denote by $\delta_x, 0 \leq \delta_x \leq g - 1$, the g-ary digits of $1/p$,

$$\frac{1}{p} = \sum_{x=1}^{\infty} \delta_x g^{-x}.$$

Since g is a primitive root modulo p, the sequence $\delta_1, \delta_2, \ldots$, is periodic with the smallest period $p - 1$.

For a divisor $s | p - 1$ and $j = 0, 1, \ldots, s - 1$, we denote

$$T_j(p, s, g) = \sum_{x=1}^{(p-1)/s} \delta_{sx+j} - \frac{(g - 1)(p - 1)}{2s}.$$

The relation between $T_j(p, s, g)$ and $h^-(p, s)$ is given by the following result which is Theorem 1 of [21]. Let s be even and let $p \equiv s + 1 \pmod{2s}$. Then

$$h^-(p, s) = \frac{2\gamma(s, p)}{g^r + 1} \prod_{\substack{\zeta \in \mathbb{C}, \\ \zeta^r = -1}} \sum_{j=1}^{r} T_j(p, s, g)\zeta^j \qquad (10.1)$$

where $r = s/2$, and

$$\gamma(s, p) = \begin{cases} p, & \text{if } p = s + 1, \\ 1, & \text{otherwise.} \end{cases}$$

We put

$$T(p, s, g) = \max_{0 \le j \le s-1} |T_j(p, s, g)| \quad \tau(p, s, g) = \frac{1}{s} \left[\sum_{j=0}^{s-1} |T_j(p, s, g)|^2 \right]^{1/2}.$$

Theorem 10.1 *The bound*

$$T(p, s, g) \ll \min\{p^{1/2}, p^{5/8}s^{-3/8}, p^{3/4}s^{-5/8}\}g \ln p$$

holds.

Proof It is easy to see that $\delta_x = d$ if and only if

$$\frac{d}{g} \le \left\{ \frac{g^{x-1}}{p} \right\} < \frac{d+1}{g}.$$

These inequalities are equivalent to $g^x \equiv y \pmod{p}$ with some y from the interval $dp/g \le y < (d+1)p/g$. Thus

$$\sum_{x=1}^{(p-1)/s} \delta_{sx+j} = \frac{1}{p} \sum_{d=0}^{g-1} d \sum_{x=1}^{(p-1)/s} \sum_{dp/g \le y < (d+1)p/g}$$
$$\times \sum_{a=-(p-1)/2}^{(p-1)/2} \mathbf{e}\left(a(g^{sx+j-1} - y)/p\right).$$

Changing the order of summation and separating the term corresponding to $a = 0$, we get

$$|T_j(p, s, g)|$$

$$\le \frac{1}{p} \left| \sum_{d=0}^{g-1} d \sum_{1 \le |a| \le (p-1)/2} \sum_{dp/g \le y < (d+1)p/g} \mathbf{e}(-ay/p) \sum_{x=1}^{(p-1)/s} \mathbf{e}(ag^{sx+j-1}/p) \right|$$

$$= \frac{1}{p} \left| \sum_{d=1}^{g-1} \sum_{1 \le |a| \le (p-1)/2} \sum_{dp/g \le y < p} \mathbf{e}(-ay/p) \sum_{x=1}^{(p-1)/s} \mathbf{e}(ag^{sx+j-1}/p) \right|$$

$$\le \frac{1}{p} \sum_{d=0}^{g-1} \sum_{1 \le |a| \le (p-1)/2} \left| \sum_{dp/g \le y < p} \mathbf{e}(ay/p) \right| \left| \sum_{x=1}^{(p-1)/s} \mathbf{e}(ag^{j-1}g^{sx}/p) \right|.$$

Applying the bounds (3.15), (3.16) and (3.17) after simple evaluations, we obtain the result. $\qquad\square$

Theorem 10.2 *The bound*

$$\tau(p, s, g) = O(gp^{1/2}s^{-1}\ln p)$$

holds.

Proof Proceeding as in the proof of Theorem 10.1, we obtain

$$\tau(p,s,g)^2$$

$$\leq \frac{1}{p^2 s^2} \sum_{j=0}^{s-1}\left[\sum_{d=0}^{g-1}\sum_{1\leq|a|\leq(p-1)/2}\left|\sum_{dp/g\leq y<p}\mathbf{e}(ay/p)\right|\right.$$

$$\left.\times\left|\sum_{x=1}^{(p-1)/s}\mathbf{e}(ag^{sx+j-1}/p)\right|\right]^2$$

$$\ll \frac{g^2}{s^2}\sum_{j=0}^{s-1}\left[\sum_{1\leq|a|\leq(p-1)/2}\frac{1}{|a|}\left|\sum_{x=1}^{(p-1)/s}\mathbf{e}(ag^{sx+j-1}/p)\right|\right]^2$$

$$= \frac{g^2}{s^2}\sum_{j=0}^{s-1}\sum_{1\leq|a|,|b|\leq(p-1)/2}\frac{1}{|a||b|}\sum_{x,y=1}^{(p-1)/s}\mathbf{e}\left((ag^{sx}-bg^{sy})g^{j-1}/p\right)$$

$$= \frac{g^2}{s^2}\sum_{1\leq|a|,|b|\leq(p-1)/2}\frac{1}{|a||b|}\sum_{x,y=1}^{(p-1)/s}\sum_{j=0}^{s-1}\mathbf{e}\left((ag^{sx}-b)g^{sy+j-1}/p\right)$$

$$= \frac{g^2}{s^2}\sum_{1\leq|a|,|b|\leq(p-1)/2}\frac{1}{|a||b|}\sum_{x=1}^{(p-1)/s}\sum_{c=1}^{p-1}\mathbf{e}\left((ag^{sx}-b)c/p\right).$$

The inner sum is equal to $p-1$ if $b \equiv ag^{sx} \pmod p$ (thus only for at most one x for each pair of a and b) and to -1 otherwise. Taking into account that the total sum is a positive number we may drop the negative terms. Therefore

$$\tau(p,s,g)^2 \ll \frac{g^2 p}{s^2}\sum_{1\leq|a|,|b|\leq(p-1)/2}\frac{1}{|a||b|} \ll g^2 ps^{-2}\ln^2 p$$

and the bound follows. $\qquad\square$

Applying the geometric-mean–arithmetic-mean inequality to (10.1), we obtain

$$\prod_{\substack{\zeta\in\mathbb{C},\\ \zeta^r=-1}}\left|\sum_{j=1}^r T_j(p,s,g)\zeta^j\right|^{2/r} \leq \frac{1}{r}\sum_{\substack{\zeta\in\mathbb{C},\\ \zeta^r=-1}}\left|\sum_{j=1}^r T_j(p,s,g)\zeta^j\right|^2$$

$$= \sum_{j=1}^r |T_j(p,s,g)|^2 = 0.5s^2\tau^2(p,s,g)$$

since it is easy to see that $T_j(p, s, g) = -T_{j+r}(p, s, g)$, $j = 1 \ldots, r$. Thus we get from (10.1)

$$h^-(p, s) \le \frac{2^{-s/4+1}\gamma(s, p)s^{s/2}\tau^{s/2}(p, s, g)}{g^{s/2} + 1}.$$

Hence, Theorem 10.2 entails that

$$h^-(p, s)^{1/s} \ll p^{1/4}s^{-1/4}\ln^{1/2} p.$$

Now we consider another question which leads to quite a different connection between cyclotomic fields and the distribution of g^x modulo p.

Let p be prime. For a divisor t of $p - 1$, G_t denotes the subgroup of $\mathbb{F}_p^*$ of size $|G_t| = t$ which is the group of sth powers of $\mathbb{F}_p^*$ where $s = (p-1)/t$. Let $R(t, p, a)$ be the number of representations $a = g_0 + \cdots + g_{s-1}$ of $a \in \mathbb{F}_p$ as a sum of elements $g_0, \ldots, g_{s-1}$ from each coset of G_t, that is the number of solutions of the equation

$$a = \sum_{j=0}^{s-1} \vartheta^{j+x_j s}, \qquad 0 \le x_1, \ldots, x_{s-1} \le t - 1,$$

where ϑ is a primitive root of $\mathbb{F}_p$. This question is studied in papers [62, 63]. It turns out that it has surprisingly many relations to a number of quite different deep problems of number theory.

In those papers, it is shown that $R(t, p, a)$ takes only two different values: $R(t, p, 0)$ and the same values for all non-zero $a \in \mathbb{F}_p$. Thus $R(t, p, 0) + (p-1)R(t, p, 1) = t^s$. Also, there is one more non-trivial relation

$$R(t, p, 0) - R(t, p, 1) = \prod_{j=0}^{s-1} \eta_j,$$

where

$$\eta_j = \sum_{x=0}^{t-1} \mathbf{e}(\vartheta^{j+xs}/p), \qquad j = 0, \ldots, s - 1,$$

are *Gaussian periods*, see [62, 63].

It is known that η_j, $j = 0, \ldots, s - 1$, are conjugate integer algebraic numbers. Therefore, from now on we may concentrate on studying the norm $\mathrm{Nm}_{\mathbb{Q}(\eta_0)/\mathbb{Q}}(\eta_0)$. Actually, it is more convenient to work with

$$N(t, p) = |\mathrm{Nm}_{\mathbb{Q}(\eta_0)/\mathbb{Q}}(\eta_0)|^{1/s}.$$

To see the connection to the cyclotomic field $\mathbb{Q}(\mathbf{e}(1/p))$, we observe that $\mathbb{Q}(\eta_0) = \mathbb{K}_s$, where $\mathbb{K}_s$ is the unique subfield of $\mathbb{Q}(\mathbf{e}(1/p))$ of degree

$[\mathbb{K}_s : \mathbb{Q}] = s$ (thus precisely the same $\mathbb{K}_s$ we discussed above) and

$$\eta_0 = \mathrm{Tr}_{\mathbb{Q}(\mathbf{e}(1/p))/\mathbb{K}_s}\,(\mathbf{e}(1/p)),$$

thus

$$N(t, p) = \left|\mathrm{Nm}_{\mathbb{K}_s/\mathbb{Q}}\left(\mathrm{Tr}_{\mathbb{Q}(\mathbf{e}(1/p))/\mathbb{K}_s}\,(\mathbf{e}(1/p))\right)\right|^{1/s}.$$

In [62] the following way to estimate $N(t, p)$ was proposed. The geometric-mean–arithmetic-mean inequality implies

$$N(t, p) \le \left(s^{-1} \sum_{j=1}^{s} |\eta_j|^2\right)^{1/2}.$$

The last sum can be computed precisely and is equal to $p - t$ (it is another version of the second identity in (3.2)) and we obtain

$$N(t, p) < t^{1/2}. \tag{10.2}$$

Also, in [63] a very nice asymptotic formula for $\ln N(t, p)$ (when $p \to \infty$) was conjectured.

Let $m = \varphi(t)$ and let $\Psi_t(X)$ be the tth cyclotomic polynomial. We define polynomials $L_i(X)$ of degree $\deg L_i < m$ from the congruences

$$L_i(X) \equiv X^i \pmod{\Psi_t(X)}, \qquad i = 0, \dots, t - 1.$$

Now for a vector $\overline{\mathbf{x}} = (x_0, \dots, x_{m-1})$, we consider linear forms $\widetilde{\mathcal{L}}_i(\overline{\mathbf{x}})$ obtained from the polynomials $L_i(X)$, $i = 0, \dots, t - 1$ by replacing X^j with x_j, $j = 0, \dots, m - 1$. Now we can define the function

$$F(x_0, \dots, x_{m-1}) = F(\overline{\mathbf{x}}) = \sum_{i=1}^{t} \mathbf{e}\left(\widetilde{\mathcal{L}}_i(\overline{\mathbf{x}})\right).$$

Then the conjecture is that, if t is a fixed prime and $p \to \infty$, then

$$\ln N(t, p) \sim \int \cdots \int_{[0,1]^m} \ln |F(\overline{\mathbf{x}})|\, d\overline{\mathbf{x}}. \tag{10.3}$$

Let us explain the very natural reasons leading to this conjecture. Put $g = \vartheta^s$. First of all we note that since $g^i \equiv L_i(g) \pmod{p}$, $i = 0, \dots, t - 1$, then

$$\eta_j = F(\vartheta^j/p, \vartheta^j g/p, \dots, \vartheta^j g^{m-1}/p), \qquad j = 0, \dots, s - 1.$$

It is evident that

$$N(t, p) \;=\; \prod_{j=0}^{s-1} F(\vartheta^j/p, \vartheta^j g/p, \dots, \vartheta^j g^{m-1}/p)^{1/s}$$

$$=\; \prod_{a=1}^{p-1} F(a/p, ag/p, \dots, ag^{m-1}/p)^{1/(p-1)}.$$

It follows from Theorem 12 [63], that the sequence of points

$$(a/p, ag/p, \dots, ag^{m-1}/p), \qquad a = 1, \dots, p-1,$$

is asymptotically (for $p \to \infty$) uniformly distributed modulo 1 in the $(m-1)$-dimensional unit cube. Moreover, a version of the bound (12.5) below and the Erdős–Turan inequality can be used to give an explicit bound for its discrepancy. In particular, for any continuous function $f(x_0, \dots, x_{m-1})$,

$$\frac{1}{p-1} \sum_{a=1}^{p-1} f(a/p, ag/p, \dots, ag^{m-1}/p)$$

$$\sim \int_0^1 \cdots \int_0^1 f(x_0, \dots, x_{m-1}) \, dx_0 \dots dx_{m-1}$$

and even the error term can be estimated. Unfortunately, the function $\ln |F(\overline{\mathbf{x}})|$ is not continuous. This is the main reason why we still cannot prove (10.3).

Nevertheless, in [63], it was proved for $t = 3$ (that is, $s = (p-1)/3$). In this case, the set of singular points of $\ln |F(x_1, x_2)|$ is finite while for $t \geq 5$, it is not. Cases of $t = 2$ and $t = 4$ are even simpler, $N(2, p) = N(4, p) = 1$. We remark that, in any case, the integral in (10.3) can be used as the upper bound for $\ln N(t, p)$ in the following form, for any prime t

$$\limsup_{p \to \infty} \ln N(t, p) \leq \int \cdots \int_{[0,1]^m} \ln |F(\overline{\mathbf{x}})| \, d\overline{\mathbf{x}}. \qquad (10.4)$$

To see this, as in [63], one may consider the continuous function

$$G_M(\overline{\mathbf{x}}) = \max \{-M, \ln |F(\overline{\mathbf{x}})|\}$$

for which we have

$$\limsup_{p \to \infty} \ln N(t, p) \;\leq\; \frac{1}{p-1} \sum_{a=1}^{p-1} \ln G_M(a/p, ag/p, \dots, ag^{m-1}/p)$$

$$\sim \int \cdots \int_{[0,1]^m} \ln |G_M(\overline{\mathbf{x}})| \, d\overline{\mathbf{x}}.$$

Taking $M \to \infty$, we obtain the required inequality.

Although at the moment it is not clear how to prove (10.3) in the full generality, below we compute the integral in the right-hand side of (10.4) which can be used as an upper bound (presumably, an approximate value) of $\ln N(t, p)$ for $t \to \infty$ and large p compared with t. We need the following statement.

Lemma 10.3. *For any $\varepsilon > 0$,*

(i) *there is a polynomial $Q_1(x)$ such that $Q_1(x) \geq \ln x$ for all $x > 0$ and*

$$\int_0^\infty (Q_1(x) - \ln x)\, e^{-x}\, dx \leq \varepsilon;$$

(ii) *there is a polynomial $Q_2(x)$ such that $Q_2(x) \geq \ln x$ for all $x > 0$ and*

$$\int_0^\infty (Q_2(x) - \ln x)\, e^{-x/2} x^{-1/2}\, dx \leq \varepsilon.$$

Proof Put $\delta = \varepsilon/5$. First of all, we remark that one can find a function $g(x)$ with first continuous derivative and such that

$$g(x) \geq \ln x + \delta e^{x/2}, \qquad \int_0^\infty |g'(x)| e^{-x}\, dx < \infty$$

and

$$\int_0^\infty g(x) e^{-x}\, dx \leq \int_0^\infty (\ln x + \delta e^{x/2}) e^{-x}\, dx + \delta;$$

hence

$$\int_0^\infty (g(x) - \ln x)\, e^{-x}\, dx \leq 3\delta.$$

Such a function $g(x)$ can be obtained from a smooth approximation from above of $\ln x + \delta e^{x/2}$ in a 'small' neighborhood of 0.

Taking into account that polynomials form a complete system with respect to the L_2-norm with weight e^{-x} on $[0, \infty)$, we see that

$$\int_0^\infty \left(P(x) - g'(x) \right)^2 e^{-x}\, dx \leq \delta^2$$

for some polynomial $P(x)$. Now we define the polynomial $Q_1(x)$ as follows

$$Q_1(x) = g(0) + \int_0^x P(u)\, du.$$

Then for any $x \in [0, \infty)$, we have

$$|Q_1(x) - g(x)| = \int_0^x |P(u) - g'(u)| \, du$$

$$\leq \left(\int_0^x (P(u) - g'(u))^2 \, e^{-u} \, du \right)^{1/2} \left(\int_0^x e^u \, du \right)^{1/2}$$

$$\leq \delta e^{x/2}.$$

Thus $Q_1(x) \geq g(x) - \delta e^{x/2} \geq \ln x$ and

$$\int_0^\infty (Q_1(x) - g(x)) \, e^{-x} \, dx \leq \delta \int_0^\infty e^{-x/2} \, dx \leq 2\delta.$$

Furthermore,

$$\int_0^\infty (Q_1(x) - \ln x) \, e^{-x} \, dx$$

$$= \int_0^\infty (Q_1(x) - g(x)) \, e^{-x} \, dx + \int_0^\infty (g(x) - \ln x) \, e^{-x} \, dx \leq 5\delta = \varepsilon.$$

Similar considerations could be used to construct $Q_2(x)$ as well. $\qquad\Box$

Theorem 10.4 *For* $t \to \infty$,

$$\int \cdots \int_{[0,1]^m} \ln |F(\overline{x})| \, d\overline{x}$$

$$\leq 0.5 \ln t - \gamma/2 + \begin{cases} o(1), & \text{if } t \text{ is odd}; \\ -0.5 \ln 2 + o(1), & \text{if } t \text{ is even}; \end{cases}$$

where γ *is the Euler constant.*

Proof First of all, we compute the moments

$$M_k(t) = \int \cdots \int_{[0,1]^m} |F(\overline{x})|^{2k} \, d\overline{x}.$$

For $k \geq 1$, we see that

$$M_k = \sum_{i_1,\ldots,i_{2k}=0}^{t-1} \int \cdots \int_{[0,1]^m} \mathbf{e} \left(\sum_{j=1}^{2k} (-1)^j \widetilde{\mathcal{L}}_{i_j}(\overline{x}) \right) d\overline{x}.$$

The integral equals 1 if and only if the linear form in the exponent vanishes identically,

$$\sum_{j=1}^{2k} (-1)^j \widetilde{\mathcal{L}}_{i_j}(\overline{x}) = 0, \qquad \overline{x} \in [0, 1]^m;$$

otherwise it equals 0. The last condition is equivalent to the polynomial identity

$$\sum_{j=1}^{2k}(-1)^j L_{i_j}(X) = 0$$

or

$$\sum_{j=1}^{2k}(-1)^j \xi^{i_j} = 0,$$

where $\xi = \mathbf{e}(1/t)$ is a root of $\Psi_t(X)$. Therefore, $M_k(t) = W_k(t)$ for $k = 1, 2, \ldots$, where $W_k(t)$ is the number of solutions of the system of equations (5.4) which is estimated in Lemma 5.2.

It is technically easier to work with $h(\overline{\mathbf{x}}) = |F(\overline{\mathbf{x}})|^2/t$. Obviously,

$$\int \cdots \int_{[0,1]^m} \ln |F(\overline{\mathbf{x}})| \, d\overline{\mathbf{x}} = 0.5 \ln t + 0.5 \int \cdots \int_{[0,1]^m} \ln h(\overline{\mathbf{x}}) \, d\overline{\mathbf{x}}$$

and

$$\int \cdots \int_{[0,1]^m} h(\overline{\mathbf{x}})^k \, d\overline{\mathbf{x}} = M_k(t) t^{-k} = W_k(t) t^{-k}, \quad k = 0, 1, \ldots, \quad (10.5)$$

where we define $W_0(t) = 1$. Let us fix some $\varepsilon > 0$ and select a polynomial $Q(x) = Q_1(x)$ for t odd and $Q(x) = Q_2(x)$ for t even, where $Q_1(x), Q_2(x)$ are defined in Lemma 10.3. Assume that $K = \deg Q$ and

$$Q(x) = c_0 + c_1 x + \cdots + c_K x^K.$$

From Lemma 10.3, the identity (10.5), and Lemma 5.2, we subsequently obtain

$$\int \cdots \int_{[0,1]^m} \ln h(\overline{\mathbf{x}}) \, d\overline{\mathbf{x}} \leq \int \cdots \int_{[0,1]^m} Q(h(\overline{\mathbf{x}})) \, d\overline{\mathbf{x}}$$

$$= \sum_{k=0}^{K} c_k W_k(t) t^{-k} = \sum_{k=0}^{K} c_k A(k) + O(t^{-1})$$

where $A(0) = 1$ and for $k \geq 1$, $A(k)$ are defined in Lemma 5.2. Now we consider two cases.

If t is odd, then $A(k) = k!$. We use the representation

$$k! = \int_0^\infty x^k e^{-x} \, dx.$$

Hence,

$$\int \cdots \int_{[0,1]^m} Q\left(h(\overline{\mathbf{x}})\right) d\overline{\mathbf{x}} = \sum_{k=0}^{K} c_k k! + O(t^{-1})$$

$$= \int_0^\infty Q(x) e^{-x} dx + O(t^{-1})$$

$$\leq \int_0^\infty e^{-x} \ln x \, dx + \varepsilon + O(t^{-1})$$

$$= -\gamma + \varepsilon + O(t^{-1}).$$

If t is even, then $A(k) = (2k-1)!!$. We use the representation

$$(2k-1)!! = (2\pi)^{-1/2} \int_0^\infty x^k e^{-x/2} x^{-1/2} \, dx.$$

Hence,

$$\int \cdots \int_{[0,1]^m} Q\left(h(\overline{\mathbf{x}})\right) d\overline{\mathbf{x}}$$

$$= \sum_{k=0}^{K} c_k (2k-1)!! + O(t^{-1})$$

$$= (2\pi)^{-1/2} \int_0^\infty Q(x) e^{-x/2} x^{-1/2} \, dx + O(t^{-1})$$

$$\leq e(2\pi)^{-1/2} \int_0^\infty e^{-x/2} x^{-1/2} \ln x \, dx + (2\pi)^{-1/2} \varepsilon + O(t^{-1})$$

$$= -\ln 2 - \gamma + (2\pi)^{-1/2} \varepsilon + O(t^{-1}).$$

Taking $\varepsilon \to 0$ and then $t \to \infty$ we obtain the statement. $\qquad\square$

The proof of Theorem 10.4 is inspired by the following probabilistic considerations. One can view $\overline{\mathbf{x}}$ as a random vector. Then $F(\overline{\mathbf{x}})$ is a complex random variable for odd t and a real random variable for even t. Roughly speaking, asymptotic expressions for moments $M_k(t)$ mean that $F(\overline{\mathbf{x}})$ can be approximated by the Gaussian complex random variable ξ with the expectation $\mathbf{E}\xi = 0$ and the dispersion $\mathbf{D}\xi = t$ for odd t and by the analogous Gaussian real random variable for even t. This makes it possible to estimate $\mathbf{E} \ln |F(\overline{\mathbf{x}})|$.

Obviously, Theorem 10.4 and the inequality (10.4) together imply

$$\limsup_{t \to \infty} \limsup_{p \to \infty} \frac{N(t, p)}{t^{1/2}} \leq \begin{cases} e^{-\gamma/2}, & \text{if } t \text{ is odd;} \\ 2^{-1/2} e^{-\gamma/2}, & \text{if } t \text{ is even.} \end{cases}$$

We remark that $e^{-\gamma/2} < 0.75$ and $2^{-1/2} e^{-\gamma/2} < 0.53$.

Now we show that more precise versions of the geometric-mean–arithmetic-mean inequality of [38] and [85] provide a slight improvement of the estimate (10.2) which holds for any t and p.

Theorem 10.5 *The bound*

$$N(t, p) \le (t - 3/4)^{1/2}$$

holds.

Proof We need the following result from [38] (it is a relaxed version of Theorem 1 of that paper). For any positive $x_1, \ldots, x_s$,

$$\left(\prod_{i=1}^{s} x_i \right)^{1/s} \le \frac{1}{s} \sum_{i=1}^{s} x_i - \frac{\Delta}{2s(s-1)}$$

where

$$\Delta = \sum_{\substack{j,k=1 \\ j \ne k}}^{s} \left(x_k^{1/2} - x_j^{1/2} \right)^2.$$

We shall apply this inequality to $x_i = |\eta_i|^2$, $i = 1, \ldots, s$. So $x_1 \ldots x_s = N(t, p)^{2s}$. Evidently the numbers

$$\left(x_k^{1/2} - x_j^{1/2} \right)^2 = (|\eta_k| - |\eta_j|)^2, \qquad 1 \le j, k \le s, \ j \ne k,$$

can be separated on groups of relatively conjugate positive integer algebraic numbers. Therefore Δ is the sum of traces of several totally real integer algebraic numbers with positive conjugates. Theorem 3 of [85] claims that the trace of every such algebraic number is greater than $1.5d$ where d is its degree. Applying this result to each of the traces in Δ, we obtain

$$\Delta > 1.5s(s - 1).$$

Therefore

$$N(t, p)^2 \le s^{-1} \sum_{j=1}^{s} |\eta_j|^2 - 3/4$$

and the estimate follows.   □

For example, we have $N(3, p) \le 1.5$, which is better than $1.732\ldots$ provided by the estimate (10.2) but of course is worse than the precise value $1.381\ldots$ given by (10.3). For $t = 5$, we obtain $N(5, p) \le 2.061\ldots$ while the bound (10.2) gives $N(5, p) \le 2.236\ldots$.

Of course, for t large compared to p, our progress here is quite marginal. Nevertheless, it shows that the bound (10.2) indeed can be improved in all ranges on t and p. Also perhaps it is possible to show that the degree of all algebraic numbers $(|\eta_k| - |\eta_j|)^2$, $1 \le j < k \le s$, are large enough for large p. Thus applying Theorem 2 of [85] in place of the used Theorem 3, one can replace $3/4$ by a larger constant $\lambda = 0.866\ldots$.

Question 10.6. *Obtain non-trivial lower bounds for $N(t, p)$.*

Part five

Applications to Pseudo-Random Number Generators

11

Prediction of Pseudo-Random Number Generators

Let $g \geq 2$ be a fixed integer and let M be a positive integer that can be written with at most L g-ary digits (that is, $M < g^L$).

It is proved in [6] that, given $k = 2L + 3$ successive digits of the g-ary expansion of $1/M$, one can find M in polynomial time $L^{O(1)}$.

Then, once again in [6], we find that, under Artin's conjecture, $k = L - 1$ digits are not enough to determine M unambiguously.

Here we prove the weaker but unconditional statement that any string of $k = \lfloor (3/37 - \varepsilon)L \rfloor$ consecutive digits provides no information about M. Roughly speaking, we see without any unproven conjectures that M may take almost any value among all prime numbers $p < g^L$.

Theorem 11.1 *For any $\varepsilon > 0$, given a string of $k = \lfloor (3/37 - \varepsilon)L \rfloor$ consecutive g-ary digits, there are at least $(1 + o(1))\pi(g^L)$ prime numbers $p < g^L$ such that the g-ary expansion of $1/p$ contains this string.*

Proof For a prime p with $\gcd(p, g) = 1$, and a sequence $d_1, \ldots, d_k$ of g-ary digits,

$$0 \leq d_i \leq g - 1, \qquad i = 1, \ldots, k,$$

we denote by $N_p(d_1, \ldots, d_k)$ the number of appearances of the string $(d_1 \ldots d_k)$ in the period of the g-ary expansion of $1/p$ (it is known that for $\gcd(g, p) = 1$, the period length of such an expansion is t_p where, as before, t_p is the multiplicative order of g modulo p).

This function was introduced and treated in [44] where the bound (3.15) was used. For our application this bound is not sharp enough, but the bounds (3.16) and (3.17) already produce some non-trivial results (although for smaller values of k than in this theorem). However, using Theorem 7.10 one can obtain stronger estimates.

We show that $N_p(d_1, \ldots, d_k) > 0$ for any string $(d_1 \ldots d_k)$ and almost all prime p.

To be more precise, denote by δ_x, $0 \le \delta_x \le g - 1$, the g-ary digits of $1/p$,

$$\frac{1}{p} = \sum_{x=1}^{\infty} \delta_x g^{-x}.$$

We observe that, for x and any g-ary string $(d_1 \ldots d_k)$, we have $\delta_{x+i} = d_i$, $i = 1, \ldots, k$, if and only if

$$\frac{D}{g^k} \le \left\{ \frac{g^x}{p} \right\} < \frac{D+1}{g^k}, \qquad (11.1)$$

where $D = d_1 g^{k-1} + d_2 g^{k-2} + \cdots + d_k$.

The inequalities (11.1) both together are equivalent to solvability of the congruence $g^x \equiv y \pmod{p}$ with some y from the interval

$$\frac{Dp}{g^k} \le y < \frac{(D+1)p}{g^k},$$

which follows from the solvability of the congruence

$$g^x \equiv b + y \pmod{p}, \qquad 0 \le y \le h,$$

where

$$b = \left\lfloor \frac{(2D+1)p}{2g^k} \right\rfloor, \qquad h = \left\lfloor \frac{p}{2g^k} \right\rfloor - 1.$$

From Lemma 9.7 we see that for all prime $p < g^L$, except possibly $o(\pi(g^L))$, we can apply Theorem 7.10 from which the desired result follows.   □

For a set of primes of positive density, the constant $3/37$ can be improved. Indeed, it follows from paper [3] that there exists an absolute constant $c > 0$ such that for sufficiently large x there are at least $c\pi(x)$ primes $p \le x$ for which $p - 1$ has a prime divisor $l \ge p^{0.677}$. We remark that in [3] this lower bound is claimed for just infinitely many primes, but, in fact, it is proved there for a set of primes of positive density. Excluding $o(\pi(x))$ of primes $p \le x$ for which the multiplicative order of g modulo p is smaller than $p^{1/2}$, see [72] or Lemma 9.7, we obtain a set of at least $(c + o(1))\pi(x)$ primes $p \le x$ for which $t_p \ge p^{0.677}$. Repeating the scheme of the proofs of Theorems 7.10 and 11.1, one can improve the value of the constant $3/37$ for primes from this remaining set.

Further, using results of [44] one can show that for an arbitrary $\varepsilon > 0$, any string of $k \le (1 - \varepsilon)L$ consecutive digits appears in the g-ary expansion of $1/M$ for at least $C(g)g^{\varepsilon L/2}$ values of $M < g^L$. Here $C(g)$ is some constant depending on g only.

Theorem 11.2 *There is a constant $c(g) > 0$, depending only on g, such that for every element M of the set*

$$\mathfrak{M} = \{M = p^\alpha \mu \ : \ 1 \le \mu \le Q, \ (\mu, g) = 1\},$$

where

$$Q = \left\lfloor c(g) g^{\varepsilon L/2} \right\rfloor,$$

p is the smallest odd prime number with $\gcd(p, g) = 1$, and p^α is the largest power of p that is less than g^L/Q, and for any string of $k = \lfloor (1 - \varepsilon)L \rfloor$ consecutive g-digits, the g-ary expansion of $1/M$ contains the string.

Proof As in the previous theorem, we denote by $N_M(d_1, \dots, d_k)$ the number of appearances of a g-ary string $(d_1, \dots, d_k)$ on the period of the g-ary expansion of $1/M$ (it is known that for $\gcd(M, g) = 1$ the period of this expansion is the multiplicative order t_M of g modulo M). It is evident that $t_M \le \varphi(M) < M$.

Also, we consider the corresponding discrepancy

$$D_M = \max_k \ \max_{(d_1, \dots, d_k)} \ |N_M(d_1, \dots, d_k) - t_M/g^k|$$

over all g-ary strings

$$(d_1, \dots, d_k), \qquad 0 \le d_i \le g - 1; \quad i = 1, \dots, k, \quad k = 1, 2, \dots.$$

Let $M = p_1^{\alpha_1} \dots p_r^{\alpha_r}$ be the prime number factorization of M. Set $m = p_1 \dots p_r$. Assuming $\gcd(M, g) = 1$ denote by $\tau_M = t_m$ the multiplicative order of g modulo m.

We need the bound

$$D_M < 2\tau_M$$

which is a relaxed version of Theorem 1 of [44].

Evidently, for any $M \in \mathfrak{M}$, $t_M \ge t_{p^\alpha}$. It is known in [44] that if we define

$$\beta = \mathrm{ord}_p(g^{p-1} - 1) < \frac{p \ln g}{\ln p}$$

then

$$t_{p^\alpha} = t_p p^{\alpha - \beta}, \qquad \alpha = \beta, \beta + 1, \dots.$$

Thus $t_M \ge C(g) g^L Q^{-1}$ where $C(g)$ is some effective constant depending on g only (since p can be estimated in terms of g). Also we evidently have the bound

$$\tau_M = t_m < m \le pQ.$$

Therefore, if $k = \lfloor (1 - \varepsilon)L \rfloor$, then for any $M \in \mathfrak{M}$ we have

$$N_M(d_1, \ldots, d_k) \geq t_M g^{-k} - D_M > C(g)g^L/Qg^k - 2pQ.$$

Putting $c(g) = (C(g)/2p)^{-1/2}$, we get that $N_M(d_1, \ldots, d_k) > 0$ for any $M \in \mathfrak{M}$ and any string $(d_1, \ldots, d_k)$. □

Since the set $\mathfrak{M}$ is exponentially large, this result means that even as many as $k = \lfloor (1 - \varepsilon)L \rfloor$ digits give us very little information on the possible values of M.

Question 11.3. *Prove that for some constant $c > 0$ and $k = \lfloor L - c \ln L \rfloor$, any (or almost any) string of k consecutive digits appears in the g-ary expansion of $1/M$ for exponentially many different values of $M < g^L$.*

Applications of results of this type to cryptography are pointed out in [6].

The problem considered here belongs to a wide class of predictability and unpredictability problems for various pseudo-random number generators, see [6], [16], and an excellent survey [46].

Here we touch on one more question from this area concerning linear congruential pseudo-random number generators. More precisely, we consider sequences of integers satisfying the conditions

$$u_n \equiv \lambda u_{n-1} \pmod{M}, \qquad 0 < u_n < M, \qquad n = 1, 2, \ldots . \quad (11.2)$$

Here the *initial value* $u_0 = a$ and the *multiplier* λ are integers which are relatively prime to *modulus* M.

For integers $s, k \geq 1$, denote by $B_{k,s}(a, \lambda, M)$ the binary sk-dimensional vector obtained by adjoining s leading bits of k first elements of the sequence (11.2). Note that we consider all elements of the sequence having

$$L = \lceil \log M \rceil$$

bits thus the vector $B_{k,s}(a, \lambda, M)$ is obtained by concatenation of bits of the s-bit integers $\lfloor u_j/2^{L-s} \rfloor$, $j = 0, \ldots, k-1$.

It is shown in [16], Theorem 3.1, that for any $k \geq 1$, $\varepsilon > 0$ and sufficiently large square-free $M > c(k, \varepsilon)$, there is an exceptional set $E(M, k, \varepsilon)$ of multipliers λ of cardinality

$$|E(M, k, \varepsilon)| \leq M^{1-\varepsilon}$$

such that for any multiplier not in $E(M, k, \varepsilon)$ the following is true. If

$$s = \lceil (1/k + \varepsilon) \log M + k(1/2 + \log 3) + 3.5 \log k + 2 - \log 3 \rceil,$$

then the entire sequence $u_0, u_1, \ldots$ can uniquely be recovered in polynomial time (per element) by knowledge of $B_{k,s}(a, \lambda, M)$.

It is also remarked in that paper that some sort of exceptional set $E(M, k, \varepsilon)$ is necessary as $\lambda = 1$ is always a 'bad' multiplier. Also it follows from Dirichlet's principle that if

$$s \leq \left\lceil \frac{1 - \varepsilon}{k} \log M \right\rceil,$$

then there is an initial value a_0 such that $B_{k,s}(a, \lambda, M) = B_{k,s}(a_0, \lambda, M)$ for at least M^ε initial values of $a = 0, \ldots, M - 1$. Thus such an initial value is secure in the following sense: the knowledge of $B_{k,s}(a_0, \lambda, M)$ is not enough to determine the entire sequence.

Here we show that for infinitely many primes $M = p$, and all but at most $p^{1-\varepsilon}$ multipliers λ, any initial value $a = 0, \ldots, p-1$ is secure in the sense that the knowledge of $B_{k,s}(a, \lambda, M)$ is not enough to determine the entire sequence, where

$$s = \left\lceil \frac{1 - \varepsilon}{k} \log p - \log \log p - 3 \right\rceil.$$

Theorem 11.4 *Let p be a sufficiently large L-bit prime with $2^L - 2^{3L/4} < p < 2^L$. Then for any $k \geq 5$, $\varepsilon > 0$, there is an exceptional set $F(p, k, \varepsilon)$ of multipliers λ of cardinality $|F(p, k, \varepsilon)| \leq p^{1-\varepsilon}$ such that for any multiplier $\lambda \notin F(p, k, \varepsilon)$, the following is true. If*

$$s \leq \frac{1 - \varepsilon}{k} L - \log L - 3,$$

then for any binary sk-dimensional vector B, there exist at least p^ε initial values of $a = 0, \ldots p - 1$ such that $B_{k,s}(a, \lambda, p) = B$.

Proof We may assume that $p > 7$. We split the vector B into s-dimensional vectors $B_0, \ldots, B_{k-1}$. Let $b_0, \ldots, b_{k-1}$ be integers whose binary representations are given by $B_0, \ldots, B_{k-1}$. We put $l = L - s$. We observe that $B_{k,s}(a, \lambda, p) = B$ if and only if

$$u_j = b_j 2^l + z_j, \quad 0 \leq z_j \leq \min\{2^l - 1, \, p - b_j 2^l - 1\}, \quad j = 0, \ldots, k-1,$$

where $u_j \equiv a\lambda^j \pmod{p}$, $0 \leq u_j \leq p - 1$.

Denote $d_j = b_j 2^l$, $h_j = \min\{2^l, \, p - d_j\}$. It follows from the condition of the theorem that

$$h_j \geq 2^l - 2^{3L/4} \geq 2^{l-1} \geq p 2^{-s-1}, \quad j = 0, \ldots, k-1, \tag{11.3}$$

provided L is large enough.

We see that the last system of equations is satisfied if and only if the following system of congruences

$$a\lambda^j \equiv x_j \pmod{p}, \qquad d_j \le x_j \le d_j + h_j - 1, \ j = 0, \dots, k-1,$$

is solvable, and thus has a unique solution. Thus the number of initial values $a = 0, \dots, p-1$ with $B_{k,s}(a, \lambda, M) = B$ is equal to the number $T(\lambda)$ of solutions of the system of congruences

$$a\lambda^j \equiv x_j \pmod{p}, \qquad j = 0, \dots, k-1,$$

where

$$d_j \le x_j \le d_j + h_j - 1 \qquad 0 \le a \le p-1.$$

On the other hand, we have

$$T(\lambda) = p^{-k} \sum_{a=0}^{p-1} \sum_{x_0=d_0}^{d_0+h_0-1} \cdots \sum_{x_{k-1}=d_{k-1}}^{d_{k-1}+h_{k-1}-1}$$

$$\times \sum_{-p/2<m_0,\dots,m_{k-1}<p/2} \mathbf{e}\left(\sum_{j=0}^{k-1} m_j(a\lambda^j - x_j)/p\right)$$

$$= h_0 \dots h_{k-1} p^{-k} + p^{-k} \sum_{\substack{-p/2<m_0,\dots,m_{k-1}<p/2 \\ m_0^2+\cdots+m_{k-1}^2>0}} \sum_{a=0}^{p-1} \mathbf{e}\left(a\sum_{j=0}^{k-1} m_j\lambda^j/p\right)$$

$$\times \prod_{j=0}^{k-1} \sum_{x_j=d_j}^{d_j+h_j-1} \mathbf{e}(m_j x_j/p).$$

For integer m, we put $\overline{m} = \max\{|m|, 1\}$ and define

$$\delta(m) = \begin{cases} 1, & \text{if } m \equiv 0 \pmod{p}, \\ 0, & \text{otherwise.} \end{cases}$$

From the identity

$$\sum_{a=0}^{p-1} \mathbf{e}(am/p) = p\delta(m)$$

and the following known inequality

$$\left| \sum_{x=d}^{d+h-1} \mathbf{e}(mx/p) \right| \le p/\overline{m},$$

which holds for any integers d, h, m with $1 \le h \le p$, $-p/2 < m < p/2$, we obtain

$$|T(\lambda) - h_0 \ldots h_{k-1} p^{-k+1}|$$

$$\le p \sum_{\substack{-p/2 < m_0, \ldots, m_{k-1} < p/2 \\ m_0^2 + \cdots + m_{k-1}^2 > 0}} \frac{\delta(m_0 + \lambda m_1 + \cdots + \lambda^{k-1} m_{k-1})}{\overline{m}_0 \ldots \overline{m}_{k-1}}.$$

Therefore for the average value

$$T = \frac{1}{p} \sum_{\lambda=1}^{p-1} |T(\lambda) - h_0 \ldots h_{k-1} p^{-k+1}|,$$

we derive

$$T \le \sum_{\substack{-p/2 < m_0, \ldots, m_{k-1} < p/2 \\ m_0^2 + \cdots + m_{k-1}^2 > 0}} \frac{1}{\overline{m}_0 \ldots \overline{m}_{k-1}}$$

$$\times \sum_{\lambda=1}^{p-1} \delta(m_0 + \lambda m_1 + \cdots + \lambda^{k-1} m_{k-1})$$

$$\le (k-1) \sum_{\substack{-p/2 < m_0, \ldots, m_{k-1} < p/2 \\ m_0^2 + \cdots + m_{k-1}^2 > 0}} \frac{1}{\overline{m}_0 \ldots \overline{m}_{k-1}}$$

$$\le (k-1) \left(1 + 2 \sum_{m=1}^{p/2} 1/m\right)^k < (3 \ln p)^k,$$

provided that p is large enough.

Let $F(p, k, \varepsilon)$ be the set of $\lambda = 1, \ldots, p-1$ for which

$$|T(\lambda) - h_0 \ldots h_{k-1} p^{-k+1}| \ge p^\varepsilon (3 \ln p)^k.$$

From the previous inequality, we derive

$$|F(p, k, \varepsilon)| < \frac{pT}{p^\varepsilon (3 \ln p)^k} \le p^{1-\varepsilon}.$$

Now for any $\lambda \notin F(p, k, \varepsilon)$, from (11.3) we obtain

$$T(\lambda) > h_0 \ldots h_{k-1} p^{-k+1} - p^\varepsilon (3 \ln p)^k > p2^{-k(s+1)} - p^\varepsilon (3 \ln p)^k$$
$$\ge p2^{-k(s+2)} \ge p^\varepsilon$$

if $s \le (1 - \varepsilon)k^{-1} \log p - \log \log p - 3$ and p large enough. $\qquad \square$

It follows from the results about the distribution of prime numbers that there are infinitely many p satisfying the condition of Theorem 11.4.

No doubt a similar result can be obtained for square-free moduli M. On the other hand, for an arbitrary M, only a weaker result can be expected (with k^2 instead of k in the denominator, see the estimate of the sum $Q_s(M)$ in Chapter 12). Also a similar result can be expected to explicit constructions of 'good' multipliers λ given in Chapter 12.

Question 11.5. *Extend the results of this chapter to other generators of pseudo-random numbers.*

12

Congruential Pseudo-Random Number Generators

For integers $a_1, \dots, a_s, M$, we set

$$
\begin{aligned}
\rho(a_1, \dots, a_s; M) &= \min \overline{m}_1 \dots \overline{m}_s, \\
\omega(a_1, \dots, a_s; M) &= \min(m_1^2 + \dots + m_s^2)^{1/2},
\end{aligned}
$$

where the minima are taken over all non-trivial solutions of the congruence

$$
a_1 m_1 + \dots + a_s m_s \equiv 0 \pmod{M},
$$

and where $\overline{m} = \max\{1, |m|\}$.

The integers $a_1, \dots, a_s$ for which $\rho(a_1, \dots, a_s; M)$ is large enough, say of order $M^{1-\varepsilon}$, are of special interest. Such integers can be used for building up formulas for approximate integration and are called *optimal coefficients* modulo M, see [43, 67, 69].

It is also known [43, 67, 69] that 'almost all' vectors $(a_1, \dots, a_s)$ are optimal coefficients for a prime modulus $M = p$ with

$$
\rho(a_1, \dots, a_s; M) \gg M/\ln^{s-1} M.
$$

Moreover, it is easy to prove that an analogous result holds for arbitrary composite M. Unfortunately, these results are non-constructive. Some algorithms for finding optimal coefficients are presented in [43]. The best for a special M has computing time $O(M^{4/3+\varepsilon})$. Recently in [45], an algorithm with computing time $O(M^{1+\varepsilon})$ was given for $M = 2^m$; the algorithm uses Hensel lifting. A different algorithm, but with the same computing time, is designed in [7]. These papers deal with arithmetical complexity, but it is not difficult to obtain the same estimates for the Boolean complexity as well.

For the case $s = 2$, we can set $a_1 = 1$; $a_2 = F_k$; $M = F_{k+1}$, where $k = 1$, $2, \dots$, and $\{F_k\}$ is the Fibonacci sequence (see [43, 67]).

The theory of diophantine approximation in $\mathbb{R}$ allows one to obtain generalizations of these results [67]. It produces $a_1, \ldots, a_s$ and M with

$$\rho(a_1, \ldots, a_s; M) \gg M^{s/2(s-1)}.$$

Now we consider linear congruential pseudo-random number generators (11.2) again. We set

$$r_s(\lambda; M) = \rho(1, \lambda, \ldots, \lambda^{s-1}; M), \quad w_s(\lambda; M) = \omega(1, \lambda, \ldots, \lambda^{s-1}; M).$$

It is known that the s-dimensional discrepancy of the sequence $\gamma_n = u_n/M$, where the sequence u_n is given by (11.2), depends drastically on $r_s(\lambda; M)$, and that its lattice structure depends on $w_s(\lambda; M)$. So, the question is: how large can values of $r_s(\lambda; M)$ and $w_s(\lambda; M)$ be and how can corresponding λ be found? We also point out one more important restriction: λ should have a large multiplicative order modulo M, say be a primitive root modulo M if $M = p^m$ is a power of odd prime number p and be congruent ± 3 modulo 8 if $M = 2^m$.

In the case of prime $M = p$, it is shown in [43] that there exists a λ with

$$r_s(\lambda, p) \gg p^{1-\varepsilon}. \tag{12.1}$$

Moreover, such λ can be chosen to be a primitive root modulo p. The method for obtaining this estimate is based on a bound for the number of zeros of polynomials modulo p and loses its power when applied to an arbitrary modulus M. In general, this method produces no better than the existence of λ with

$$r_s(\lambda, M) \gg M^{1/(s-1)-\varepsilon}. \tag{12.2}$$

It is technically easier to work with the function

$$\sigma_s(\lambda, M) = \sum_{\substack{-M/2 < m_1, \ldots, m_s \leq M/2 \\ m_1^2 + \cdots + m_s^2 > 0}} \frac{\delta_M(m_1 + \lambda m_2 + \cdots + \lambda^{s-1} m_s)}{\overline{m}_1 \ldots \overline{m}_s},$$

where $\delta_M(a) = 1$ if $a \equiv 0 \pmod{M}$ and $\delta_M(a) = 0$ otherwise. Moreover as we could see in the proof of Theorem 11.4, the discrepancy of the corresponding sequence is determined by this function rather than by $r_s(\lambda, M)$. On the other hand, the following relations between $r_s(\lambda, M)$ and $\sigma_s(\lambda, M)$ (which can be found in [67, 69] for example) show that they are each quite close to the other:

$$r_s(\lambda, M)^{-1} \leq \sigma_s(\lambda, M) \ll r_s(\lambda, M)^{-1}(\ln M)^s.$$

Both these estimates (12.1) and (12.2) can be obtained by considering the sum

$$Q_s(M) = \sum_{\substack{\lambda=1 \\ \gcd(\lambda,M)=1}}^{M} \sigma_s(\lambda, M)$$

$$= \sum_{\substack{-M/2<m_1,\dots,m_s\leq M/2 \\ m_1^2+\cdots+m_s^2>0}} \frac{1}{\overline{m}_1 \dots \overline{m}_s}$$

$$\times \sum_{\substack{\lambda=1 \\ \gcd(\lambda,M)=1}}^{M} \delta_M(m_1 + \lambda m_2 + \cdots + \lambda^{s-1}m_s).$$

The inner sum does not exceed the number of solutions of the non-trivial polynomial congruence

$$m_1 + \lambda m_2 + \cdots + \lambda^{s-1}m_s \equiv 0 \pmod{M}, \qquad 1 \leq \lambda \leq M$$

which can be nicely estimated as $s - 1$ for prime $M = p$ (as it is done in the proof of Theorem 11.4) and only as $O(M^{1-1/(s-1)})$ for arbitrary M (see [40, 42]). These two estimates lead to (12.1) and (12.2) respectively.

Moreover, we show that for arbitrary integer M, this way cannot lead to a better estimate of $r_s(\lambda, M)$ because

$$Q_s(M) \geq 2^{-(s-1)(s-2)} M^{1-1/(s-1)}$$

for an infinite sequence of M. Indeed, let $M = 2^m$ and let $m = k(s-1)$. Then, it is easy to see that for $\lambda = 2^k l - 1$ with any $l = 1, \dots, 2^{k(s-2)}$ we have

$$\sum_{i=0}^{s-1} \binom{s-1}{i} \lambda^i = (\lambda+1)^{s-1} \equiv 0 \pmod{M}.$$

Hence,

$$\sigma_s(\lambda, M) \geq r_s(\lambda, M)^{-1} \geq 2^{-(s-1)(s-2)}$$

for at least $2^{k(s-2)} = M^{1-1/(s-1)}$ values of λ with $1 \leq \lambda \leq M$, $\gcd(\lambda, M) = 1$, and the assertion follows.

The case $s = 2$ can be dealt with by using the results on two-dimensional optimal coefficients (we may set $\lambda \equiv a_1/a_2 \pmod{M}$).

The first non-trivial case is $s = 3$. This case is considered in [47] for fixed prime power moduli M (these are the most interesting moduli). There is shown that for $M = 2^m$, there exists a $\lambda \equiv 5 \pmod{8}$ with

$$r_3(\lambda, M) \gg M/\ln^2 M.$$

For an arbitrary s, some non-trivial lower bounds can be obtained as well. For instance, if we define λ by the congruence $\lambda\tau \equiv \vartheta \pmod{M}$, with $\gcd(\vartheta\tau, M) = 1$ and $\vartheta \sim \tau \sim M^{1/(s+1)}$, then

$$r_s(\lambda, M) \gg M^{2/(s+1)}. \tag{12.3}$$

If in the previous construction one takes $\vartheta \sim \tau \sim (M/2s)^{1/s}$, then

$$w_s(\lambda, M) \geq 2^{1/2}(2s)^{-1/2s} M^{1/s} + o(M^{1/s}). \tag{12.4}$$

It also can be shown that for any λ of a prime multiplicative order $t = s + 1$ modulo a prime p, the lower bound

$$w_s(\lambda, p) \geq (s + 1)^{-(s-1)/2s} p^{1/s} \tag{12.5}$$

holds. Indeed, suppose that

$$m_1 + \lambda m_2 X + \cdots + \lambda^{s-1} m_s \equiv 0 \pmod{p}$$

and not all $m_1, \ldots, m_s$ are zeros. Because $\lambda^{s+1} \equiv 1 \pmod{p}$, the resultant $R = \operatorname{Res}(F, G)$ of the polynomials

$$F(X) = m_1 + m_2 X + \cdots + m_s X^{s-1} \quad \text{and} \quad G(X) = 1 + X + \cdots + X^s$$

is divisible by p. Because $t = s + 1$ is prime, $G(X)$ is irreducible, therefore $R \neq 0$ and thus $|R| \geq p$. On the other hand, by the Hadamard inequality,

$$|R|^2 \leq \left(\sum_{i=1}^{s} m_i^2\right)^s (s + 1)^{s-1}$$

and (12.5) follows.

We note that we use the same considerations to derive the bounds (14.6) and (14.15).

Surprisingly (12.4) and (12.5) are of the same order (at least for s fixed) and both meet the upper bound $w_s(\lambda, M) \leq \gamma_s M^{1/s}$, where γ_s is the Hermite constant (see Section 3.3.4 of [37]).

Similar problems also arise in cryptography. Let $W_s(\delta, M)$ be the size of the set of λ, $1 \leq \lambda \leq M$, with $w_s(\lambda, M) \leq M^\delta$. In [16], among other very interesting and important results, for $s = 3$, the bound

$$W_3(\delta, M) = O(M^{1/2+3\delta/2+\varepsilon})$$

is proved.

For the case that is most important for applications, when $M = 2^m$, the bound (12.3) was improved in Section 10.1 of [84]. Here we consider a slightly more general case when $M = p^m$ is a power of a fixed prime number p.

Theorem 12.1 *Let p be a fixed prime and let ϑ, $1 \le \vartheta \le p^2$ be a primitive root modulo p^2 if $p \ge 3$ and $\vartheta = 3$ if $p = 2$. Set*

$$\lambda \equiv (p^t + \vartheta)(p^t + 1)^{-1} \quad (\bmod \ p^m), \tag{12.6}$$

where $t = 2\lfloor m/(2s + 1) \rfloor$. Then, for sufficiently large $M = p^m$,

$$r_s(\lambda, M) \gg M^{4/(2s+1)}$$

and the multiplicative order λ modulo M is at least $M/4$.

Proof Assume that $m \ge 4s + 2$, thus $t \ge 4$. Then $\lambda \equiv \vartheta \ (\bmod \ p^3)$ and λ is a primitive root modulo M if $p \ge 3$ (thus of multiplicative order $(p-1)p^{m-1} \ge M/4$) or $\lambda \equiv 3 \ (\bmod\ 8)$ if $p = 2$ (thus of multiplicative order $2^{m-2} = M/4$). In order to estimate $r_s(\lambda, M)$, we prove the following general statement. Let the integers μ, η, M satisfy the conditions

- $\mu \sim \eta \sim M^{2/(2s+1)}$, $\mu \ne \eta$, $|\mu - \eta| = O(1)$;
- $M \equiv k \ (\bmod \ \mu)$, $M \equiv l \ (\bmod \ \eta)$ with some k and l such that $M^{1/(2s+1)} \ll |k|, |l| \ll M^{1/(2s+1)}$;
- $\gcd(\mu, M) = \gcd(\eta, M) = 1$;

and let $\lambda \equiv \mu\eta^{-1} \ (\bmod \ M)$. Then $r_s(\lambda, M) \gg M^{4/(2s+1)}$. It is easy to verify that $M = p^m$, $\mu = p^t + \vartheta$, $\eta = p^t + 1$ satisfy these conditions. Indeed, only the second condition is not obvious. To verify it, we put

$$\tau = m - st - t/2 = m - (2s + 1)\lfloor m/(2s + 1) \rfloor,$$

and note that $0 \le \tau \le 2s$. One sees that

$$M \equiv (-\vartheta)^s p^{t/2+\tau} \quad (\bmod \ \mu)$$
$$M \equiv (-1)^s p^{t/2+\tau} \quad (\bmod \ \mu),$$

and that

$$M^{1/(2s+1)} \ll \left|(-1)^s p^{t/2+\tau}\right| \le \left|(-\vartheta)^s p^{t/2+\tau}\right| \ll M^{1/(2s+1)}.$$

Suppose that

$$m_1 + \lambda m_2 + \cdots + \lambda^{s-1} m_s \equiv 0 \quad (\bmod \ M).$$

From the definition of λ for some integer r, we obtain

$$m_1 \mu^{s-1} + m_2 \mu^{s-2}\eta + \cdots + m_s \eta^{s-1} = rM. \tag{12.7}$$

Let $\delta = \gcd(\mu, \eta)$. We have

$$\delta = \gcd(\mu - \eta, \eta) \leq |\mu - \eta| = O(1).$$

If $r = 0$, then it is clear that the first non-zero m_i is divisible by η/δ and the last non-zero m_j is divisible by η/δ, $1 \leq i < j \leq s$. Therefore

$$r_s(\lambda, M) \geq \mu \eta \delta^{-2} \gg M^{4/(2s+1)}.$$

From here on we assume that $r \neq 0$. Without loss of generality we can suppose that $|m_1| \leq |m_s|$. We consider two cases:

I. If $|m_1| \geq 0.5 \, M \mu^{-s+1}$, then we immediately obtain

$$r_s(\lambda, M) \geq \overline{m}_1 \overline{m}_s \gg M^2 \mu^{-2s+2} \gg M^{4/(2s+1)}.$$

II. If $|m_1| < 0.5 \, M \mu^{-s+1}$, then $|rM - m_1 \mu^{s-1}| \geq |r|M/2$. Hence,

$$
\begin{aligned}
\overline{m}_2 \ldots \overline{m}_s \; &\gg \; \overline{m}_2 + \cdots + \overline{m}_s \\
&\gg \; M^{-2(s-1)/(2s+1)} |m_2 \mu^{s-2} \eta + \cdots + m_s \eta^{s-1}| \\
&= \; M^{-2(s-1)/(2s+1)} |rM - m_1 \mu^{s-1}| \gg M^{3/(2s+1)} |r|.
\end{aligned}
$$

Therefore,

$$r_s(\lambda, M) \gg M^{3/(2s+1)} \overline{m}_1 |r|. \tag{12.8}$$

It follows from (12.7) that

$$m_1 \mu^{s-1} - rM \equiv 0 \quad (\mathrm{mod}\ \eta).$$

Let $\gamma = \mu - \eta$, then $m_1 \gamma^{s-1} - rl \equiv 0 \ (\mathrm{mod}\ \eta)$. Again we consider two cases.

II.1. If $m_1 \gamma^{s-1} - rl = 0$, then

$$|m_1| = |rl\gamma^{-s+1}| \gg |l| \gg M^{1/(2s+1)}.$$

II.2. If $m_1 \gamma^{s-1} - rl \neq 0$, then $|m_1 \gamma^{s-1} - rl| \geq \eta$ and

$$\overline{m}_1 |r| \geq |m_1| + |r| - 1 \geq |(m_1 \gamma^{s-1} - rl)/l\gamma^{s-1}| - 1 \gg \eta/|l| \gg M^{1/(2s+1)}.$$

Therefore in each case we have $\overline{m}_1 |r| \gg M^{1/(2s+1)}$. Plugging this inequality into the estimate (12.8), we obtain the theorem. $\qquad\square$

Question 12.2. *Can the constructions giving (12.3), (12.4), and (12.6) be modified to produce λ for which they hold for all s up to some bound S (say, can one find λ with $r_2(\lambda, 2^m) \geq 2^{4m/5}$ and $r_3(\lambda, 2^m) \geq 2^{4m/7}$)?*

Question 12.3. *Obtain an upper bound of $W_s(\delta, M)$ for $s > 3$.*

A similar question can be asked about $R_s(\delta, M)$ which is defined as the size of the set of λ, $1 \le \lambda \le M$, with $r_s(\lambda, M) \le M^\delta$. Actually, the bound (12.1) was proved by showing that, for a prime p, $R_s(\delta, M) = o(M)$ for any $\delta < 1$, while for an arbitrary integer M, the best we can do is to show that $R_s(\delta, M) = o(M)$ for any $\delta < 1/(s-1)$ (which leads to (12.2)).

Question 12.4. *Does the bound $R_s(\delta, M) = o(M)$ hold for any $s \ge 2$ and $\delta < 1$?*

Question 12.5. *Can the theory of diophantine approximation in p-adic fields be used to get better lower bounds and constructions?*

Indeed, let $M = p^m$. Then, in the p-adic metric,

$$\left| m_1 + m_2\lambda + \cdots + m_s\lambda^{s-1} \right| > p^{-m}$$

for $\overline{m}_1, \ldots, \overline{m}_s < r_s(\lambda, M)$. Roughly speaking, this means that the λ we are looking for should not be well p-adically approximated by roots of polynomials with small integer coefficients.

Part six

Applications to Finite Fields

13

Small mth Roots Modulo p

Let p be an odd prime, and let $m > 1$ be a divisor of $p - 1$. For each, an mth power residue a modulo p, $\rho_{m,p}(a)$, is defined as the largest absolute value of all m solutions of the congruence

$$x^m \equiv a \pmod{p}, \qquad -(p-1)/2 \leq x \leq (p-1)/2.$$

Clearly, $\rho_{m,p}(a)$ may take at most $(p-1)/m$ distinct values (as there are $(p-1)/m$ distinct mth power residues modulo p), which we enumerate in the non-decreasing order

$$M_1(m, p) \leq \cdots \leq M_{(p-1)/m}(m, p).$$

The question about the distribution of numbers $M_i(m, p)$ was posed in [75] where, among many other interesting results, the following upper bound has been obtained,

$$M_1(m, p) \leq 2^{\nu(m)} p^{1-1/\varphi(m)}$$

where $\nu(m)$ is the number of distinct odd prime divisors of m.

In [73] this bound is improved as follows:

$$M_1(m, p) \leq \min \left\{ p^{1-1/m}, \prod_{\substack{l \mid m \\ l \in \mathcal{P}, \, l \geq 3}} l^{1/(2l-2)} p^{1-1/\varphi(m)} \right\} \qquad (13.1)$$

where $\mathcal{P}$ denotes the set of prime numbers.

It is useful to note that the well-known asymptotic formula, see [74], Theorem 3.1 of Chapter 1, or [91], Exercise 9b of Chapter 2,

$$\sum_{l \in \mathcal{P}, l \leq N} \frac{\ln l}{l} = \ln N + O(1)$$

implies that

$$\prod_{\substack{l|m \\ l\in\mathcal{P},\, l\geq 3}} l^{1/(2l-2)} \ll (\nu(m)\,\mathrm{Ln}\,\nu(m))^{1/2} \ll \ln^{1/2} m.$$

Thus (13.1) improves drastically the former estimate of [75] with respect to the factor depending on m. For m of some special structure, better bounds are obtained, say for $m = 2^s 3^t$ we have

$$M_1(m, p) \leq 2 \times 3^{-1/2} p^{1-1/\varphi(m)}.$$

Moreover, that paper [73] provides a lower bound of the same order (at least for m fixed). In a somewhat relaxed form, it can be written down as follows:

$$M_1(m, p) \geq \min\left\{ m^{-1}p,\ \varphi(m)^{1/2}m^{-1} \prod_{\substack{l|m \\ l\in\mathcal{P},\, l\geq 3}} l^{1/(l-1)} p^{1-1/\varphi(m)} \right\}. \quad (13.2)$$

It is useful to note that

$$\varphi(m)^{1/2}m^{-1} \prod_{\substack{l|m \\ l\in\mathcal{P},\, l\geq 3}} l^{1/(l-1)}$$

$$\geq (2m)^{-1/2} \prod_{\substack{l|m \\ l\in\mathcal{P},\, l\geq 3}} (1 - 1/l)^{1/2} l^{1/(l-1)} \gg m^{-1/2}.$$

Here, using quite a different method, we obtain several more lower bounds for $M_1(m, p)$ which are mainly useful for larger values of m. In fact, paper [73] considers $M_1(m, p)$ at about its expected order $p^{1-1/\varphi(m)}$, while we are more interested in when $M_1(m, p)$ reaches its trivial upper bound $(p - 1)/2$. It is interesting to note that for m of order $\ln p$, both methods give very similar results.

First of all, we remark that there is a close relation between $M_1(m, p)$ and the function $H_p(m)$ which we studied in Chapter 7, namely

$$M_1(m, p) \geq \frac{p - 1 - H_p(m)}{2}. \quad (13.3)$$

As we mentioned in Chapter 4, studying $M_1(m, p)$ is closely related to bounds of short character sums with exponential functions.

Indeed, it is easy to verify that one can take $B = M_1(t, p)^2$ in the proof of Theorem 4.2 which now produces the estimate

$$\max_{\gcd(a,p)=1} \left| \sum_{x=1}^{t} \mathbf{e}(ag^x/p) \right| \leq t - \frac{4(t-1)M_1(t, p)^2}{p^2}.$$

Now one can get another version of the bound (4.5) simply combining the previous inequality with the lower bound (13.2) of $M_1(t, p)$ from [73].

On the other hand, there is a trivial lower bound

$$\max_{\gcd(a,p)=1} \left| \sum_{x=1}^{t} e(ag^x/p) \right| \geq t \cos(2\pi M_1(t, p)/p) = t + O(tM_1(t, p)^2 p^{-2})$$

which follows from the inequality

$$\left| \sum_{x=1}^{t} e(ag^x/p) \right| \geq \sum_{x=1}^{t} \cos(2\pi ag^x/p).$$

Now, using the upper bound (13.1) of $M_1(t, p)$ from [73], or even previous slightly weaker bounds from [75], one sees that the bound (4.5) is quite sharp.

We also remark that the question on the lower bound of $M_1(m, p)$ is in some sense dual to the question about estimating $T(l)$ which is considered in Chapter 14.

We begin our study of $M_j(m, p)$ with a simple observation that the estimates

$$mj/2 \leq M_j(m, p) \leq p/2 - (p - jm - 1)/2m \qquad (13.4)$$

follow from elementary counting arguments. Indeed, since there are exactly m different values x, $|x| \leq M_j(m, p)$ corresponding to each $M_j(m, p)$, then $2M_j(m, p) \geq mj$. Similar considerations can be used to find the upper bound.

Now we show that the method which we use in this book allows us to get sharper bounds on $M_j(m, p)$.

First of all, we estimate $M_j(m, p)$ for any j.

Theorem 13.1 *For all $j = 1, \ldots, (p-1)/m$, the bound*

$$M_j(m, p) \geq (p - 1)/2 - \frac{p^{3/2}}{mj^{1/2}}$$

holds.

Proof We fix a primitive root ϑ modulo p and set $s = (p - 1)/2$, $h = s - M_j(m, p)$, $g = \vartheta^{(p-1)/m}$.

It follows from the definition of $M_j(m, p)$ that the congruence

$$\xi g^i \equiv s + u - v \pmod{p}, \qquad 0 \leq i \leq m - 1, \quad 0 \leq u, v \leq h - 1,$$

Applications to Finite Fields

has no solution for at least mj distinct values of ξ in the interval $-(p-1)/2 \le \xi \le (p-1)/2$. For such ξ, we have

$$
\begin{aligned}
0 &= \frac{1}{p} \sum_{i=0}^{m-1} \sum_{u,v=0}^{h-1} \sum_{\lambda=0}^{p-1} \mathbf{e}\left(\lambda(\xi g^i - s - u + v)/p\right) \\
&= \frac{mh^2}{p} + \frac{1}{p} \sum_{\lambda=1}^{p-1} \mathbf{e}(-\lambda s/p) \sum_{i=0}^{m-1} \mathbf{e}(\lambda \xi g^i/p) \left|\sum_{u=0}^{h-1} \mathbf{e}(\lambda u/p)\right|^2.
\end{aligned}
$$

Hence,

$$
mh^2 \le \left|\sum_{\lambda=1}^{p-1} \mathbf{e}(-\lambda s/p) \sum_{i=0}^{m-1} \mathbf{e}(\lambda \xi g^i/p) \left|\sum_{u=0}^{h-1} \mathbf{e}(\lambda u/p)\right|^2\right|
$$

for at least mj distinct values of ξ in the interval $-(p-1)/2 \le \xi \le (p-1)/2$.

Therefore,

$$
\begin{aligned}
mj(mh^2)^2 &\le \sum_{\xi=-(p-1)/2}^{(p-1)/2} \left|\sum_{\lambda=1}^{p-1} \mathbf{e}(-\lambda s/p) \sum_{i=0}^{m-1} \mathbf{e}(\lambda \xi g^i/p) \left|\sum_{u=0}^{h-1} \mathbf{e}(\lambda u/p)\right|^2\right|^2 \\
&= \sum_{\xi=-(p-1)/2}^{(p-1)/2} \sum_{\lambda,\mu=1}^{p-1} \mathbf{e}((\mu-\lambda)s/p) \sum_{i,k=0}^{m-1} \mathbf{e}\left(\xi(\lambda g^i - \mu g^k)/p\right) \\
&\quad \times \left|\sum_{u,v=0}^{h-1} \mathbf{e}((\lambda u - \mu v)/p)\right|^2 \\
&= \sum_{\lambda,\mu=1}^{p-1} \mathbf{e}((\mu-\lambda)s/p) \left|\sum_{u,v=0}^{h-1} \mathbf{e}((\lambda u - \mu v)/p)\right|^2 \\
&\quad \times \sum_{i,k=0}^{m-1} \sum_{\xi=-(p-1)/2}^{(p-1)/2} \mathbf{e}\left(\xi(\lambda g^i - \mu g^k)/p\right).
\end{aligned}
$$

The sum over ξ is equal to $pN(\lambda, \mu)$ where $N(\lambda, \mu)$ is the number of solutions of the congruence

$$
\lambda g^i - \mu g^k \equiv 0 \pmod{p}, \qquad 0 \le i, k \le m-1.
$$

Obviously $N(\lambda, \mu) \le m$. Hence

$$mj(mh^2)^2 \le mp \sum_{\lambda,\mu=1}^{p-1} \mathbf{e}((\mu - \lambda)s/p) \left|\sum_{u,v=0}^{h-1} \mathbf{e}((\lambda u - \mu v)/p)\right|^2$$

$$\le mp \left[\sum_{\lambda=1}^{p-1} \left|\sum_{u=0}^{h-1} \mathbf{e}(\lambda u/p)\right|^2\right]^2 \le mp^3h^2.$$

Therefore, $h \le p^{3/2}/mj^{1/2}$ and the estimate follows. □

Note that (13.4) implies that $M_j(m, p) \sim (p-1)/2$ if $jmp^{-1} \sim 1$, while Theorem 13.1 does the same for $jm^2p^{-1} \to \infty$.

In particular, for $j = 1$, we obtain $M_1(m, p) > (p-1)/2 - p^{3/2}m^{-1}$. Here we show how to estimate $M_1(m, p)$ using the bound (3.16) and (3.17).

Theorem 13.2 *The bound*

$$M_1(m, p) > (p-1)/2 + O\left(\min\left\{m^{-5/8}p^{5/4}, m^{-3/8}p^{9/8}\right\}\right)$$

holds.

Proof Let a be an integer with $\rho_{m,p}(a) = M_1(m, p)$.

We fix a primitive root ϑ modulo p and set $g = \vartheta^{(p-1)/m}$.

If ξ is a root of the congruence $x^m \equiv a \pmod{p}$, then other roots are of the form $x_i = \xi g^i$, $i = 0, \ldots, m-1$.

Let $s = (p-1)/2$, $h = s - M_1(m, p)$.

It follows from the definition of a that the congruence

$$\xi g^i \equiv s + u - v \pmod{p}, \qquad 0 \le i \le m-1, \quad 0 \le u, v \le h-1,$$

has no solution. On the other hand, this number can be expressed as

$$0 = p^{-1} \sum_{i=0}^{m-1} \sum_{u,v=0}^{h-1} \sum_{\lambda=0}^{p-1} \mathbf{e}\left(\lambda(\xi g^i - s - u + v)/p\right)$$

$$= mh^2p^{-1} + p^{-1} \sum_{\lambda=1}^{p-1} \mathbf{e}(-\lambda s/p) \sum_{i=0}^{m-1} \mathbf{e}(\lambda \xi g^i/p) \left|\sum_{u=0}^{h-1} \mathbf{e}(\lambda u/p)\right|^2.$$

Applying the bounds (3.16) and (3.17), we obtain

$$mh^2 \leq \sum_{\lambda=1}^{p-1} \left| \sum_{i=0}^{m-1} \mathbf{e}(\lambda \xi g^i / p) \right| \left| \sum_{u=0}^{h-1} \mathbf{e}(\lambda u / p) \right|^2$$

$$\ll \min\{m^{3/8} p^{1/4}, m^{5/8} p^{1/8}\} \sum_{\lambda=1}^{p-1} \left| \sum_{u=0}^{h-1} \mathbf{e}(\lambda u / p) \right|^2$$

$$\ll h \min\{m^{3/8} p^{5/4}, m^{5/8} p^{9/8}\}.$$

Therefore

$$h \ll \min\left\{ m^{-5/8} p^{5/4}, \ m^{-3/8} p^{9/8} \right\}$$

and the estimate follows. $\square$

In particular,

$$M_j(m, p) \sim (p-1)/2, \qquad j = 1, \ldots, (p-1)/m,$$

if $mp^{-1/3} \to \infty$.

In some cases, better estimates can be obtained from (13.3). For example, if $m \geq p^{1/2}$ then Theorem 7.10, combined with the bound (13.3), implies that

$$M_j(m, p) = (p-1)/2 + O(p^{34/37+\varepsilon}), \qquad j = 1, \ldots, (p-1)/m.$$

A slightly weaker estimate which nevertheless holds for very small values of m is provided by Lemma 6 of [41].

Theorem 13.3 *For $m > \ln p (\mathrm{Ln} \ln p)^{-1+\varepsilon}$, the bound*

$$M_1(m, p) \gg p/(\ln p)^{1/2+\varepsilon}$$

holds.

Theorem 13.3 is quite precise. Indeed it follows from (13.1) that for any $\delta > 0$, the bound $M_1(m, p) \gg p/(\ln p)^\delta$ implies

$$m \gg \ln p (\mathrm{Ln} \ln p)^{-1}.$$

On the other hand, our knowledge about m for which $M_1(m, p)$ is of order p is rather poor.

Denote by $\mu(p)$ the smallest μ such that $M_1(m, p) > (p-1)/4$ if $m \geq \mu$, $p \equiv 1 \pmod{m}$. The bound (13.1) and Theorem 13.2 imply

$$\limsup_{p \to \infty} \mu(p)/\ln p > 0 \quad \text{and} \quad \limsup_{p \to \infty} \mu(p)/p^{3/8} < \infty,$$

respectively.

Question 13.4. *Obtain tighter upper and lower bounds of $\mu(p)$.*

14

Supersingular Hyperelliptic Curves

For prime l, let $T(l)$ be the largest t with the property that there exists some integer g, $\gcd(g, l) = 1$, and with multiplicative order t modulo l and such that for some a, $\gcd(a, l) = 1$, all the smallest positive residues of ag^x $(\mod l)$, $x = 0, \ldots, t - 1$, belong to the interval $[1, (l - 1)/2]$.

The matter is related to the question considered in [92, 93] (and several other papers) on supersingular over $\mathbb{F}_p$ hyperelliptic curves of the form $y^2 = x^l + \lambda$, $\lambda \neq 0$, where l is an odd prime. It follows from Lemma 2 of [93] that if this curve is not supersingular and $p > (l - 1)/2$, then the multiplicative order of p modulo l does not exceed $T(l)$.

The first upper bound of $T(l)$ of the form

$$T(l) = O(l^{1/2} \ln l)$$

was stated in [68]. This result is improved in [81, 82, 84] as $T(l) \leq 100 l^{3/7}$. Here, using the bound (3.17), we obtain an essentially better upper bound. Its proof as well as the proofs of other similar bounds below rely on the following statement.

Lemma 14.1. *If t is such that there are integers g of multiplicative order t modulo l and a with $\gcd(ag, l) = 1$ and such that all the residues of ag^x modulo l, $x = 0, \ldots, t - 1$, belong to the interval of $[1, (l - 1)/2]$, then*

$$t \leq 3S(l, t)$$

where

$$S(l, t) = \max_{1 \leq \alpha \leq l-1} \left| \sum_{x=1}^{t} \mathbf{e}(\alpha g^x / l) \right|.$$

Proof Put $h = \lfloor l/4 \rfloor$, $b = (l + 1)/2 + h$. Then the congruence

$$ag^x \equiv b + y - z \pmod{l}, \qquad 0 \leq x \leq t - 1, \ 0 \leq y, z \leq h$$

is unsolvable. Therefore

$$0 = \frac{1}{l} \sum_{x=0}^{t-1} \sum_{y,z=0}^{h} \sum_{\alpha=0}^{l-1} \mathbf{e}\left(\alpha(ag^x - b - y + z)/l\right).$$

Changing the order of summation and separating the term $t(h+1)^2/l$ corresponding to $\alpha = 0$, we obtain from (3.4)

$$
\begin{aligned}
t(h+1)^2 & \leq \sum_{\alpha=1}^{l-1} \left| \sum_{x=0}^{t-1} \sum_{y,z=0}^{h} \mathbf{e}\left(\alpha(ag^x - b - y + z)/l\right) \right| \\
& = \sum_{\alpha=1}^{l-1} \left| \sum_{x=0}^{t-1} \mathbf{e}(\alpha ag^x/l) \right| \left| \sum_{y=0}^{h} \mathbf{e}(\alpha y/l) \right|^2 \\
& \leq S(l,t) \sum_{\alpha=1}^{l-1} \left| \sum_{y=0}^{h} \mathbf{e}(\alpha y/l) \right|^2 \\
& = S(l,t) \left(\sum_{\alpha=0}^{l-1} \left| \sum_{y=0}^{h} \mathbf{e}(\alpha y/l) \right|^2 - (h+1)^2 \right) \\
& = S(l,t) \left(l(h+1) - (h+1)^2 \right).
\end{aligned}
$$

Taking into account that $h + 1 \geq l/4$, we obtain the statement. □

Lemma 14.1 and the bound (3.17) imply

Theorem 14.2 *The bound*

$$T(l) \ll l^{1/3}$$

holds.

We also remark that Theorem 1 of [93] essentially claims that $T(l)$ is odd.

Using Theorem 14.2, it is easy to show that, for fixed l, the density $\delta(l)$ of primes p for which the corresponding curve is not supersingular over $\mathbb{F}_p$ is quite small.

Theorem 14.3 *The bound*

$$\delta(l) = O(l^{-2/3+\varepsilon})$$

holds.

More precisely,

$$\delta(l) \le (l-1)^{-1} \sum_{\substack{d|l-1, \\ d \le T(l)}} \varphi(d) = O\left(T(l)l^{-1+\varepsilon}\right).$$

It is interesting to note here that for the Fermat curve $x^m + y^m = 1$, the situation is quite different. Let $\Delta(m)$ be the density of primes moduli p for which that curve is not supersingular. An explicit formula for $\Delta(m)$ is known; it depends on the arithmetic structure of m and is quite complicated. However, for prime numbers $m = l$, it simplifies as $\Delta(l) = 2^{-k}$ where k is the largest power of 2 dividing $l - 1$. Thus for almost all l, the density of primes p such that the curve is not supersingular modulo p is not too small, in particular it is $1/2$ for a 'half' of the primes l.

For these and several other results see [76], where it is also shown that 'on average' over all integral numbers m, the density $\Delta(m)$ is very close to 1

$$\sum_{m \le M} |1 - \Delta(m)| \sim \alpha M (\ln M)^{-2/3},$$

where $\alpha > 0$ is an explicitly given absolute constant.

In fact, the previous bound of $T(l)$ from [68] also produces a non-trivial bound for $\delta(l)$. On the other hand, in the following dual question it cannot give anything because we do need something below the 'square-root' bound for $T(l)$.

Given some prime p and integer $L < 2p + 1$, let $N_p(L)$ be the number of primes $l \le L$ which correspond to non-supersingular curves. Then from Theorem 14.2, we derive the following result.

Theorem 14.4 *For $L < 2p + 1$, the bound*

$$N_p(L) \ll \frac{L^{2/3} \ln p}{\ln(L \ln p)}$$

holds.

Proof Let $V(L)$ be the maximal value of $T(l)$ over all odd primes $l \le L$. From Theorem 14.2 we see that $V(L) \ll L^{1/3}$. Furthermore, all the primes $l \le L$ corresponding to non-supersingular curves must divide the product

$$W_p(L) = \prod_{t=1}^{V(L)} (p^t - 1) = \exp\left(O(V(L)^2 \ln p)\right).$$

Thus

$$N_p(L) \le \omega\left(W_p(L)\right) \ll \frac{\ln W_p(L)}{\operatorname{Ln} \ln W_p(L)}$$

and we get the desired result. □

In particular, the curve $y^2 = x^l + \lambda$, $\lambda \in \mathbb{F}_p^*$ is supersingular over $\mathbb{F}_p$ for almost all prime numbers l of the segment $[1, L]$ with

$$L > \ln^3 p \ln \ln p.$$

This is plain because $N_p(L) = o(\pi(L))$ for such values of L. Below we improve this result and show that the same holds under a weaker condition (14.2).

It is easy to see that there exists some absolute constant $c > 0$ such that if $l - 1 = 2r_1 r_2$, where $r_1, r_2 \ge cl^{1/3}$ are prime numbers (or if $l - 1 = 2r$ where r is prime that is widely believed to be true for infinitely many prime l), then $T(l) = 1$.

Unfortunately, at the moment we cannot hope to prove that one of the representations above holds infinitely often. The current best result of [32] gives the representation $l = 2r + 1$ or $l = 2r_1 r_2 + 1$ with $r_1, r_2 \ge l^\alpha$ for some $\alpha > 1/4$ (one can take $\alpha = 0.276\ldots$, see the proof of Lemma 1, of [32]). Below we show how to use this result together with Theorem 5.5 to prove that $T(l) = 1$ infinitely often.

Theorem 14.5 *We have,*

$$\liminf_{l \to \infty} T(l) = 1.$$

Proof We shall use Lemma 1 of [32] in the following (weakened) form: for any sufficiently large x, there are at least $L(x) \gg x/\ln^2 x$ primes $l \le x$ such that either $l = 2r_1 r_2 + 1$ where $r_1, r_2 \ge x^{1/4}$ are prime numbers or $l = 2r_0 + 1$ where r_0 is a prime number. It is enough to show that $T(l) = 1$ for at least one such l.

We see that if l is large enough and $\min\{r_1, r_2\} > l^{3/8}$ or if $l = 2r_0 + 1$, then $T(l) = 1$ because of Theorem 14.2 (since $T(l)$ is odd).

Thus we may assume that all $L(x)$ prime numbers l described above have $l = 2r_1 r_2 + 1$ and $\min\{r_1, r_2\} \le l^{3/8}$. Thus there is a prime r such that $x^{1/4} \le r \le l^{3/8}$ and $r = \min\{r_1, r_2\}$ for at least

$$\frac{L(x)}{x^{3/8}} \gg x^{5/8} \ln^{-2} x$$

primes $l = 2r_1 r_2 + 1 \le x$. Now let us select $k = 3$ and $U = x^{5/8}/\ln^3 x$. in Theorem 5.5. Then for any x large enough, we see that there is a prime l such that $l = 2r_1r_2 + 1$, $r_1 \leq r_2$, are prime, $l^{3/8} \geq r_1 \gg x^{1/4}$ and

$$
\max_{\gcd(\alpha,l)=1} \left| \sum_{j=0}^{r_1-1} \mathbf{e}(\alpha g_l^{j(l-1)/r_1}/l) \right| \ll r_1 l^{1/18} \left(r_1^{-1/3} + x^{-5/72+\varepsilon} \right)
$$

$$
\leq r_1 \left(r_i^{-1/3} x l^{1/18} + l^{-1/72+\varepsilon} \right)
$$

$$
\ll r_1^{1-1/3+4/18} + r_1^{1-8/216+8\varepsilon/3}
$$

$$
\ll r_1^{26/27+8\varepsilon/3},
$$

for a sufficiently small $\varepsilon > 0$, where g_l is a primitive root modulo l.

Using Lemma 14.1, we obtain that $T(l) \neq r_1$.

We also see that $r_2 = (l-1)/2r_1 \gg l^{5/8}$. Thus from Theorem 14.2 we conclude that $T(l) \neq r_2$ either, provided that x is sufficiently large. Recalling that $T(l)$ is odd, we see that $T(l) = 1$. $\qquad\square$

In fact, it is easy to see that analogously one can prove that $T(l) \ll l^{\varepsilon}$ for almost all primes l. This is a simple corollary of the bound

$$
\sum_{l \leq L} T(l)^{1/2} \ll L^{1+\varepsilon}. \tag{14.1}
$$

Without loss of generality, we may assume that $\varepsilon < 1/2$. Let us fix some $t \geq L^{\varepsilon}$. Applying Theorem 5.5 with $k = \lfloor 0.5\varepsilon^{-1} \rfloor + 1$ and $U = L^{1/2+\varepsilon}$, we see that for all primes $l \leq L$ except maybe at most $O(L^{1/2+\varepsilon})$ of them, we obtain a non-trivial bound (of the form $o(t)$) of the exponential sums corresponding to all integers g with multiplicative order t modulo l. Therefore Lemma 14.1 implies that $T(l) = t$ is possible only for those exceptional primes. Thus,

$$
|\{l \leq L : T(l) = t\}| = O(L^{1/2+\varepsilon}).
$$

Recalling the bound of Theorem 14.2 and taking the sum over all t, we see that, for some absolute constant $C > 0$,

$$
\sum_{l \leq L} T(l)^{1/2} = \sum_{1 \leq t \leq CL^{1/3}} t^{1/2} |\{l \leq L : T(l) = t\}|
$$

$$
\ll (L^{\varepsilon})^{1/2} L + \sum_{L^{\varepsilon} \leq t \leq CL^{1/3}} t^{1/2} L^{1/2+\varepsilon}
$$

$$
\ll L^{1+\varepsilon/2} + L^{1+\varepsilon},
$$

and the bound (14.1) follows.

Now we show that the curve $y^2 = x^l + \lambda$, $\lambda \in \mathbb{F}_p^*$ is supersingular over $\mathbb{F}_p$ for almost all prime numbers l of the segment $[1, L]$ with

$$L > \ln^{1+\varepsilon} p \qquad (14.2)$$

which improves the previously mentioned result following from Theorem 14.4.

Indeed, for almost all primes $l \leq L$, we have $T(l) \leq L^{\varepsilon/3}$. Denote

$$W_{p,\varepsilon}(L) = \prod_{1 \leq t \leq L^\varepsilon} (p^t - 1).$$

Then we have

$$N_p(L) \leq o(\pi(L)) + \omega\left(W_{p,\varepsilon/3}(L)\right) = o(\pi(L)).$$

Question 14.6. *Is it true that $T(l) = O(\ln l)$?*

Now we consider the question about lower bounds of $T(l)$. In particular, we show that the upper bound of Question 14.6 is actually the best possible (if it is true, of course).

It is shown in [64] that $T(l) \geq 3$ infinitely often. Furthermore, in [64] a construction was proposed showing that $T(l) \geq r$ whenever l is a prime of the form $l = (a^r - 1)/(a - 1)$ for any integer $a > 2$ and odd $r > 1$ thus $T(l) \gg \ln l$ for such l. Certainly, we cannot prove that there are infinitely many such l. Nevertheless, below we prove that

$$\limsup_{l \to \infty} \frac{T(l)}{\ln l} \geq \frac{1}{\ln(2 + \pi/2)}. \qquad (14.3)$$

For real z, we denote by $\|z\|$ the distance from z to the nearest integer.

Theorem 14.7 *Let $h \in (0, 1/2)$. A 1-periodic function $\psi(t)$ such that $\psi(t) = 0$ for $\|t\| \geq h$ has the Fourier expansion*

$$\psi(t) = \sum_{u=-\infty}^{+\infty} d_u \mathbf{e}(ut),$$

with coefficients satisfying the inequalities

$$1 = d_0 > |d_u|, \qquad |d_u| \ll u^{-2}, \qquad (14.4)$$

for all integers $u \neq 0$ and the identity

$$\sum_{u=-\infty}^{+\infty} |d_u| = B. \qquad (14.5)$$

Let r be a sufficiently large prime. For every integer l with $l \equiv 1 \pmod{r}$, we fix an integer g_l of multiplicative order r modulo l. Then, for all except at most

$O(B^r)$ *primes* $l \equiv 1 \pmod{r}$, *for any real* $t_0, \ldots, t_{r-1}$ *there exists an integer* a, *with* $\gcd(a, l) = 1$ *such that the inequalities*

$$\|a g_l^j / l - t_j\| < h, \qquad j = 0, \ldots, r-1,$$

hold.

Proof Let $\mathcal{L}$ be an arbitrary set of $N = |\mathcal{L}|$ primes $l \equiv 1 \pmod{r}$. We will show that for at least one $l \in \mathcal{L}$ and a with $\gcd(a, l) = 1$, the inequalities of the theorem hold for $j = 0, \ldots, r-1$, provided that N is large enough. For an integer vector $\overline{u} = (u_0, \ldots, u_{r-1}) \in \mathbb{Z}^r$, we denote

$$\mathcal{L}(\overline{u}) = \{l \in \mathcal{L} : u_0 + u_1 g_l + \cdots + u_{r-1} g_l^{r-1} \equiv 0 \pmod{l}\},$$
$$N(\overline{u}) = |\mathcal{L}(\overline{u})|,$$
$$\rho(\overline{u}) = \left(r \sum_{k=0}^{r-1} u_k^2 \right)^{1/2}.$$

Let us define the numbers u_j for all integers j by the equality $u_j = u_i$ where $i \equiv j \pmod{l}$, $0 \le i < l$. For an integer k, set

$$\overline{u}^{(k)} = (u_0, u_k, \ldots, u_{k(r-1)}) \in \mathbb{Z}^r.$$

Denote by D the diagonal $D = \{(u, \ldots, u) \in \mathbb{Z}^r\}$. Obviously, $\mathcal{L}(\overline{u}) = \mathcal{L}$ for $\overline{u} \in D$. Now we show that

$$\sum_{k=1}^{r-1} N(\overline{u}^{(k)}) \ll r \ln \rho(\overline{u}) / \ln (r \ln \rho(\overline{u})) \tag{14.6}$$

for $\overline{u} \notin D$.

Indeed, if $l \in \mathcal{L}(\overline{u}^{(k)})$ for some $k \in \{1, \ldots, r-1\}$; then the congruence

$$u_0 + u_1 g_l^k + \cdots + u_{r-1} g_l^{k(r-1)} \equiv 0 \pmod{l}$$

holds. Hence, if $|\{k : l \in \mathcal{L}(\overline{u}^{(k)})\}| = s$, then the polynomials $F(X) = u_0 + u_1 X + \cdots + u_{r-1} X^{r-1}$ and $G(X) = 1 + X + \cdots + X^{r-1}$ have at least s common zeros modulo l. Therefore, by Lemma 5.3 their resultant $R = \text{Res}(F, G)$ is divisible by l^s. Moreover, because G is an irreducible polynomial and $\overline{u} \notin D$, we obtain that $R \ne 0$. Thus,

$$\sum_{k=1}^{r-1} N(\overline{u}^{(k)}) \le \omega(|R|) \ll \ln |R| / \text{Ln} \ln |R|.$$

On the other hand, the Hadamard inequality gives

$$|R|^2 \le r^{r-1} \left(\sum_{k=0}^{r-1} u_k^2 \right)^r \le \rho(\overline{\mathbf{u}})^{2r},$$

and (14.6) follows.

Now let $\psi_k(t) = \psi(t - t_k)$ for $k = 0, \ldots, r - 1$, then

$$\psi_k(t) = \sum_{u=-\infty}^{+\infty} c_u^{(k)} \mathbf{e}(ut),$$

where $c_u^{(k)} = d_u \mathbf{e}(-ut_k)$. Clearly,

$$|c_u^{(k)}| = |d_u| \quad (u \in \mathbb{Z}; \; k = 0, \ldots, r - 1). \tag{14.7}$$

Notice that $\psi_k(t) \ne 0$ only if $\|t - t_k\| < h$. We define the quantities

$$Q_{l,a} = \prod_{k=0}^{r-1} \psi_k(ag_l^k/l), \qquad Q = \sum_{l \in \mathcal{L}} \frac{1}{l} \sum_{a=0}^{l-1} Q_{l,a}.$$

Obviously, if

$$Q - \sum_{l \in \mathcal{L}} \frac{1}{l} Q_{l,0} \ne 0,$$

then there exist $l \in \mathcal{L}$ and $a \in \mathbb{Z}$ such that $Q_{l,a} \ne 0$ and therefore the inequalities of the theorem hold for this pair of l and a. So, it is enough to prove that

$$|Q| \le \sum_{l \in \mathcal{L}} \frac{1}{l} |Q_{l,0}|$$

implies

$$N = O(B^r). \tag{14.8}$$

For $\overline{\mathbf{u}} \in \mathbb{Z}^r$, denote

$$c_{\overline{\mathbf{u}}} = \prod_{k=0}^{r-1} c_{u_k}^{(k)}.$$

We have

$$\frac{1}{l} \sum_{a=0}^{l-1} Q_{l,a} = \sum_{\overline{\mathbf{u}} : l \in \mathcal{L}(\overline{\mathbf{u}})} c_{\overline{\mathbf{u}}}.$$

We put $K = r^4$ and define the sums $\sigma_1, \sigma_2, \sigma_3$ as follows

$$\sigma_1 = \sum_{\bar{u} \in D} N(\bar{u}) c_{\bar{u}},$$

$$\sigma_2 = \sum_{\bar{u} \notin D, \rho(\bar{u}) \leq K} N(\bar{u}) c_{\bar{u}},$$

$$\sigma_3 = \sum_{\bar{u} \notin D, \rho(\bar{u}) > K} N(\bar{u}) c_{\bar{u}}.$$

We have

$$Q = \sum_{\bar{u} \in \mathbb{Z}^r} N(\bar{u}) c_{\bar{u}} = \sigma_1 + \sigma_2 + \sigma_3. \tag{14.9}$$

Let us consider these three sums separately.

Note that (14.4), for sufficiently large r, implies

$$\sum_{u \neq 0} |d_u|^r < 1/2. \tag{14.10}$$

For the first sum, from (14.7) and (14.10), we see that

$$\sigma_1 = N \left| \sum_{u \in \mathbb{Z}} \prod_{k=0}^{r-1} c_u^{(k)} \right| \geq N \left(|d_0|^r - \sum_{u \neq 0} |d_u|^r \right) > N/2. \tag{14.11}$$

If $u \notin D$ and $\rho(\bar{u}) \leq K$; then using (14.6) and recalling the choice of K, we obtain

$$\sum_{k=1}^{r-1} N(\bar{u}^{(k)}) \ll r \ln K / \ln(r \ln K) \ll r.$$

Hence, using (14.7) and (14.5), we derive

$$(r-1)\sigma_2 \ll r \sum_{\bar{u} \in \mathbb{Z}^r} |d_{\bar{u}}| \ll r \left(\sum_{u \in \mathbb{Z}} |d_u| \right)^r = B^r r. \tag{14.12}$$

To estimate the third sum, we note that $\rho(\bar{u}) \leq r|u_k|$ for at least one $k \in \{0, \dots, r-1\}$, thus (14.6) implies the inequality

$$N(\bar{u}) \ll r \ln(r|u_k|).$$

Therefore

$$\sigma_3 \ll r \sum_{k=0}^{r-1} \sum_{\substack{|u| > Kr^{-1} \\ u_k = u}} \sum_{\bar{u} \notin D} |c_{\bar{u}}| \ln(r|u|)$$

$$\leq r^2 \sum_{|u| > Kr^{-1}} |d_u| \ln(r|u|) \left(\sum_{v \in \mathbb{Z}} |d_v| \right)^{r-1}.$$

Applying (14.5) again, we obtain

$$\sigma_3 \ll B^{r-1} r^2 \sum_{|u|>r^3} |u|^{-2} \ln |u| \ll B^r. \tag{14.13}$$

Substituting (14.11), (14.12), and (14.13) in (14.9), we obtain

$$|Q| \geq N/2 + O(B^r).$$

Finally,

$$\left| \sum_{l \in \mathcal{L}} \frac{1}{l} Q_{l,0} \right| \leq B^r \sum_{j=1}^{N} \frac{1}{jr+1} < (1 + \ln N) B^r / r.$$

Thus the condition $Q - \sum_{l \in \mathcal{L}} \frac{1}{l} Q_{l,0} = 0$ implies

$$N/2 \leq (1 + \ln N) B^r / r + O(B^r),$$

and the bound (14.8) holds. □

One can verify, see [2], that the following function

$$\psi(t) = \begin{cases} \pi \cos 2\pi t & \text{for } \|t\| \leq 1/4; \\ 0 & \text{otherwise;} \end{cases}$$

satisfies the condition of Theorem 14.7. Thus we obtain the following statement.

Theorem 14.8 *Let r be a sufficiently large prime. Then, for all except at most $O\left((2 + \pi/2)^r\right)$ primes $l \equiv 1 \pmod{r}$, $T(l) \geq r$.*

We note that $2 + \pi/2 = 3.570\ldots$.

The known results about distribution of prime numbers in arithmetic progressions imply that l can be selected such that $l \equiv 1 \pmod{r}$, $T(l) \geq r$, and $l = (2 + \pi/2)^{r+O(1)}$. Thus, the inequality (14.3) holds.

The following result shows that selecting l slightly greater (of order $r^{2r+o(r)}$) we may eliminate the set of exceptional primes.

Theorem 14.9 *Let $r \geq 7$ and $l \equiv 1 \pmod{r}$ be prime numbers such that*

$$l > (200\,000r + 2)^{r-1} r^{(2r-3)/2}$$

then $T(l) \geq r$.

Proof Let g be any integer number of multiplicative order r modulo l. Put $U = 200\,000r + 2$. As in the proof of the previous theorem, estimating the corresponding resultant $R = \text{Res}(F, G)$ of polynomials $F(X) = u_0 + u_1 X +$

$\cdots + u_{r-2} X^{r-2}$ and $G(X) = 1 + X + \cdots + X^{r-1}$, we obtain that all non-trivial solutions of the congruence

$$u_0 + u_1 g + \cdots + u_{r-2} g^{r-2} \equiv 0 \pmod{l} \tag{14.14}$$

satisfy

$$\max_{k=0,\ldots,r-2} |u_k| \geq U. \tag{14.15}$$

Indeed, because r is prime, $R \neq 0$ therefore $|R| \geq l$ for any solution of the congruence (14.14). On the other hand, by the Hadamard inequality,

$$|R| \leq \left(\max_{k=0,\ldots,r-2} |u_k| \right)^{r-1} r^{(2r-3)/2}$$

and (14.15) follows.

Again, for real z, we denote by $\|z\|$ the distance from z to the nearest integer. Let us consider the polynomials

$$
\begin{aligned}
P_1(u) &= \left(1 - 0.5(u - 0.98)^2 \right)^{100\,000r}, \\
P_2(u) &= \int_{-1}^{u} P_1(v)\, dv, \\
P_3(u) &= 25 P_2(u)/P_2(1).
\end{aligned}
$$

Clearly, $P_3(-1) = 0$, $P_3(1) = 25$, and P_3 increases on $[-1, 1]$. For $u \in [-1, 0.97]$ and $u \in [0.99, 1]$ we have

$$P_1(u) \leq (1 - 5 \times 10^{-5})^{100\,000r} < e^{-5r}$$

and for $v \in [0.979, 0.981]$,

$$P_1(v) \geq (1 - 5 \times 10^{-7})^{100\,000r} > e^{-0.06r}.$$

Therefore, $P_1(v)/P_1(u) > e^{4.9r}$ for such u and v. Further,

$$\int_{-1}^{0.97} P_1(u)\, du + \int_{0.99}^{1} P_1(u)\, du < 1000 e^{-4.9r} \int_{0.979}^{0.981} P_1(v)\, dv$$

$$< 1000 e^{-4.9r} \int_{-1}^{1} P_1(u)\, du = 1000 e^{-4.9r} P_2(1)$$

and

$$P_3(0.97) < 25\,000 e^{-4.9r}, \quad 25 - P_3(0.99) < 25\,000 e^{-4.9r}.$$

Thus, for $r \geq 7$, we have $P_3(u) < 25^{-r}$ for $-1 \leq u \leq 0.97$ and $24 \leq P_3(u) \leq 25$ for $u \in [0.99, 1]$.

Consider the trigonometric polynomial $Q(t) = P_3(\cos(2\pi(t - 1/8)))$. We have

$$0 \le Q(t) \le \begin{cases} 25, & \text{for any } t; \\ 25^{-r}, & \text{for } \|t - 1/8\| > 1/24. \end{cases} \tag{14.16}$$

We define the quantities

$$Q_A = \prod_{k=0}^{r-2} Q(Ag^k/l).$$

Since all non-trivial solutions of the congruence (14.14) satisfy (14.15) and $U > \deg Q$, we have

$$(1/l) \sum_{A=0}^{l-1} Q_A = \left(\int_0^1 Q(t)\, dt \right)^{r-1}.$$

Taking into account that $24 \le Q(t) \le 25$ for $\|t - 1/8\| \le 1/48$, we deduce

$$\int_0^1 Q(t)\, dt > \int_{1/8-1/48}^{1/8+1/48} Q(t)\, dt \ge 1.$$

Hence, $Q_A > 1$ for some $A = 1, \ldots, l - 1$. From (14.16), it follows that

$$\|Ag^k/l - 1/8\| \le 1/24, \qquad k = 0, \ldots, r - 2$$

for this A.

Now, it is easy to see that for at least one of the following values of a, $a = A$, $a = 2A$, $a = 3A$ and $a = (1 + g + g^2)A$, all the residues of ag^k, $k = 0, 1, \ldots, r - 1$, belong to the interval $[0, (l - 1)/2]$. Indeed, assume that it is not true for $A, 2A$, and $3A$. Because of the choice of A, the only possible value of k for which the residues of ag^k can be outside the interval $[0, (l - 1)/2]$ is $k = r - 1$. Obviously, the residue of Ag^{r-1} modulo p belongs to the interval $[l - l/6, l - 1]$. For $a = (1 + g + g^2)A$, we have $ag^k = Ag^k + Ag^{k+1} + Ag^{k+2}$. If neither of $k, k + 1, k + 2$ is congruent to $r - 1$ modulo r, then the residue of the last expression is evidently less than $3l(1/8 + 1/24) = l/2$. If $k = r - 3, r - 2, r - 1$, then the residue of the last expression is between $2l(1/8 - 1/24) - l/6 = 0$ and $2l(1/8 + 1/24) = l/3$. $\qquad \square$

Let us mention that the idea used to derive (14.6) and (14.15) is a quantitative version of quite similar considerations of the proof of Theorem 12 of [63] (see also Chapter 10 of this book).

Certainly, Theorems 14.8 and 14.9 lead to a non-trivial lower bound of $T(l)$ for almost all primes l. More precisely, applying a sieve method, one can prove that for any function $\psi(l) \to 0$, $T(l) \ge (\ln l)^{\psi(l)}$ for almost all primes l. We remind you that we have seen that $T(l) \le l^\varepsilon$ for almost all primes l.

We believe that both these bounds can be sharpened.

Question 14.10. *Is it true that there exist two constants A_1, A_2 such that for any $\varepsilon > 0$, $(\ln l)^{A_1 - \varepsilon} \le T(l) \le (\ln l)^{A_2 + \varepsilon}$ for almost all prime l?*

It is quite possible that one can take $A_1 = A_2 = 1$.

It is interesting to find a fast algorithm to compute $T(l)$. Trivially, one can compute $T(l)$ in time $O(l^{1+\varepsilon})$. It is enough, for each divisor $t \mid l - 1$, to check if for some $u = 0, \ldots, s - 1$, all residues g^{u+sv}, $v = 0, 1, \ldots, t$, where $s = (l - 1)/t$, belong to the interval $[1, (l-1)/2]$.

Question 14.11. *Find an algorithm to compute $T(l)$ in time $O(l^{1-\delta})$ with some $\delta > 0$.*

In fact very similar results can be obtained for a more general function $T_{\alpha,\beta}(l)$ which is defined as follows.

Let α and β be real numbers, $0 < \beta - \alpha < 1$. For a prime l, let $T_{\alpha,\beta}(l)$ be the largest r with the property that there exists some integer g, $\gcd(g, l) = 1$, with multiplicative order t modulo l and such that for some a, $\gcd(a, l) = 1$ and for any $x = 0, \ldots, t - 1$, there exists $\gamma \in (\alpha, \beta)$ such that $ag^x \equiv \gamma l \pmod{l}$.

Thus

$$T_{0,1/2}(l) = T(l).$$

Generalizations of Theorems 14.2 and 14.5 are very straightforward. One can show that for any fixed α and β, we have

$$T_{\alpha,\beta}(l) = O(l^{1/3}), \qquad \liminf_{l \to \infty} T_{\alpha,\beta}(l) = 1.$$

Theorem 14.7 is formulated and proved for the general case as well. The function

$$\psi(t) = \max \left\{ \frac{h - \|t\|}{h^2}, 0 \right\}$$

shows that for any $h < 1/2$, we can take $B = 1/h$ in Theorem 14.7. Thus, taking $h = (\beta - \alpha)/2$ we obtain

$$\limsup_{l \to \infty} \frac{T_{\alpha,\beta}(l)}{\ln l} \ge \frac{1}{1 - \ln(\beta - \alpha)}.$$

However we remark that this general construction does not produce the bound (14.3). Certainly, finding a general construction coinciding with (14.3) for $h = 1/4$ would be of great interest.

In fact, the class of functions of Theorem 14.7 can be extended if one uses some standard 'smoothing' technique from [34].

For a 1-periodic function ψ, we denote by supp ψ the closure of the set $\{t \in [-1/2, 1/2] : \psi(t) \neq 0\}$.

Let $0 < h < 1/2$ and let

$$B(h) = \inf_{\text{supp } \psi \subset [-h,h]} \left\{ \sum_{u=-\infty}^{+\infty} |d_u| \right\},$$

where infimum is taken over all continuous 1-periodic functions ψ such that supp $\psi \subset [-h, h]$, and ψ has the Fourier expansion

$$\psi(t) = \sum_{u=-\infty}^{+\infty} d_u \mathbf{e}(ut),$$

with $d_0 = 1$.

Theorem 14.12 *For any α, β with $0 < \beta - \alpha < 1$, the bound*

$$\limsup_{l \to \infty} \frac{T_{\alpha,\beta}(l)}{\ln l} \geq \frac{1}{\ln B \left((\beta - \alpha)/2 \right)} \tag{14.17}$$

holds.

Proof The proof is based on Theorem 14.7.

Fix an arbitrary $\varepsilon \in (0, 1)$. By the definition of $B(h)$, there exists a continuous 1-periodic function ψ_1 such that supp $\psi_1 \subset [-h, h]$, ψ_1 has the Fourier expansion

$$\psi_1(t) = \sum_{u=-\infty}^{+\infty} d_u^{(1)} \mathbf{e}(ut),$$

with $d_0^{(1)} = 1$ and

$$\sum_{u=-\infty}^{+\infty} |d_u^{(1)}| \leq B(h) + \varepsilon.$$

Let $0 < \delta < \min(h/2, 1/2 - h)$. Following [34], we define the function

$$V_\delta(t) = 2\Delta_{2\delta}(t) - \Delta_\delta(t),$$

where

$$\Delta_\delta(t) = \max\{1 - \|t\|/\delta, 0\}.$$

Let

$$\psi_2(t) = \psi_1(t) \left(1 - V_\delta(t - h) - V_\delta(t + h) \right),$$

and

$$\psi_2(t) = \sum_{u=-\infty}^{+\infty} d_u^{(2)} \mathbf{e}(ut).$$

We have

$$\operatorname{supp} \psi_2 \subset [-h+\delta, h-\delta]. \qquad (14.18)$$

It is known ([34], Chapter 5) that for any sufficiently small δ,

$$\sum_{u=-\infty}^{+\infty} |d_u^{(2)} - d_u^{(1)}| < \varepsilon.$$

So,

$$|d_0^{(2)}| > 1 - \varepsilon, \quad \sum_{u=-\infty}^{+\infty} |d_u^{(2)}| < B(h) + 2\varepsilon. \qquad (14.19)$$

Let

$$D = \max_u |d_u^{(2)}|, \quad S = \max\{u \; : \; |d_u^{(3)}| = D\},$$

$$\psi_3(t) = \psi_2(t)\mathbf{e}(-St)/d_S^{(2)}.$$

By (14.18) and (14.19), we get

$$\operatorname{supp} \psi_3 \subset [-h+\delta, h-\delta] \qquad (14.20)$$

and

$$\sum_{u=-\infty}^{+\infty} |d_u^{(2)}| < \frac{B(h) + 2\varepsilon}{1 - \varepsilon}. \qquad (14.21)$$

Also, by the choice of D and S,

$$d_0^{(3)} = 1 \geq |d_u| \qquad (14.22)$$

for all integers u.
Now define

$$\psi(t) = \int_{-1/2}^{1/2} \psi_3(s)\Delta_\delta(t-s)/\delta.$$

By (14.20),

$$\operatorname{supp} \psi \subset [-h, h]. \qquad (14.23)$$

Let

$$\Delta_\delta(t)/\delta = \sum_{u=-\infty}^{+\infty} c_u \mathbf{e}(ut),$$

and

$$\psi(t) = \sum_{u=-\infty}^{+\infty} d_u \mathbf{e}(ut).$$

Then, for any $u \neq 0$, we have the inequalities

$$1 = c_0 > |c_u|, \quad |c_u| \ll u^{-2},$$

and, for any u, we also have

$$d_u = d_u^{(3)} c_u.$$

Therefore, taking into account (14.21) and (14.22), we find that the function ψ satisfies the conditions of Theorem 14.7 with

$$B < \frac{B(h) + 2\varepsilon}{1 - \varepsilon}.$$

Now we can use that theorem for $t_j = (\alpha + \beta)/2$, $j = 1, \ldots, r$, and $h = (\beta - \alpha)/2$.

The known results about distribution of prime numbers in arithmetic progressions imply that l can be selected such that $l \equiv 1 \pmod{r}$, $T_{\alpha,\beta}(l) \geq r$, and

$$l = B^{r+O(1)} < \left(\frac{B(h) + 2\varepsilon}{1 - \varepsilon} \right)^{r+O(1)}.$$

Since ε is arbitrarily small, the inequality (14.17) holds.   □

Unfortunately, we do not know how to extend Theorem 14.9 to the case of $T_{\alpha,\beta}(l)$.

We finish this chapter by mentioning that the problem we considered in this chapter is a discrete version of the still unsolved 3/2-problem of Mahler [55] about the existence of $\eta > 0$ such that fractional parts $\{\eta(3/2)^x\} < 1/2$, $x = 1, 2, \ldots$. An exhaustive survey of previously known results as well as many new ones can be found in [15, 87].

15

Distribution of Powers of Primitive Roots

Let ϑ be a fixed primitive root modulo p. For $1 \le a \le p - 1$, denote by $T_{a,\vartheta}(N, M, H)$ the number of solutions of the congruence

$$a\vartheta^x \equiv M + u \pmod{p}, \qquad x = 1, \ldots, N, \quad u = 0, \ldots, H - 1.$$

It is easy to see that the Burgess bound of character sums (see Chapter 6 of [52] for a survey including this and related bounds) immediately infers that there is an absolute positive constant c such that

$$T_{a,\vartheta}(N, M, H) = NH/p + O\left(H^{1-c\varepsilon^2}\right)$$

for any $H > p^{1/4+\varepsilon}$.

Then it was proved in [57] that

$$T_{a,\vartheta}(N, M, H) = NH/p + O(p^{1/2} \ln^2 p).$$

In fact this bound had been known for a long time (see [44, 45, 52, 67, 69, 84]). It can easily be derived from the following known bound of exponential sums that

$$\max_{1 \le a \le p-1} \left| \sum_{x=1}^{N} \mathbf{e}(a\vartheta^x/p) \right| \ll p^{1/2} \ln p. \tag{15.1}$$

However, in [57], a new method is proposed providing new upper bounds for the average value

$$R_{a,\vartheta}(N, H) = \frac{1}{p} \sum_{M=0}^{p-1} \left(T_{a,\vartheta}(N, M, H) - NH/p\right)^2.$$

The following bounds are obtained in that paper,

$$R_{a,\vartheta}(N, H) = \frac{NH}{p} \left(1 + O(Hp^{-1} + Np^{-1} + p^{1/2}HN^{-1} \ln^3 p)\right)$$

131

and

$$R_{a,\vartheta}(N, H) = \frac{NH}{p}\left(1 + O(Hp^{-1} + Np^{-1} + p^{1/4+\varepsilon}HN^{-3/4}\right.$$
$$\left. + p^{5/8+\varepsilon}N^{-7/8} + p^{3/8+\varepsilon}HN^{-1})\right)$$

which is stronger than the first one for some values of parameters. Of course these bounds are much stronger than the considerably more easily proved bound

$$R_{a,\vartheta}(N, H) = O(H \ln p) \tag{15.2}$$

holding for every $N \leq p - 1$ and $H \leq p$. It immediately follows from the identity (15.6) below if one substitutes the bound (15.1) in it.

These results are motivated by applications to some sorting algorithm, namely to the QuickSort algorithm using a congruential pseudo-random number generator [36, 90]. For this application, only the case $HN \sim p$ is really important. In fact, for the application mentioned, we need only $H = \lfloor p/N \rfloor$. For such N and H, the first bound is worse than the second one which (together with (15.2) for $N \leq p^{3/8}$) produces

$$R_{a,\vartheta}(N, \lfloor p/N \rfloor) = \begin{cases} 1 + o(1), & \text{if } N \geq p^{5/7+\varepsilon}; \\ O(p^{5/4+\varepsilon}N^{-7/4}), & \text{if } p^{5/7+\varepsilon} \geq N \geq p^{1/2}; \\ O(p^{11/8+\varepsilon}N^{-2}), & \text{if } p^{1/2} \geq N \geq p^{3/8}; \\ O(p^{1+\varepsilon}N^{-1}), & \text{if } p^{3/8} \geq N. \end{cases} \tag{15.3}$$

It is very plausible that the first asymptotic in (15.3) holds for a wider range of N. In particular, it is mentioned in [57] that one may expect the same result for $p^{\varepsilon} \leq N \leq p - 1$. On the other hand, perhaps it is not an easy matter as even the very strong Generalized Lindelöf Hypothesis would imply only the exponent $2/3$ in place of $5/7$, see [57].

Actually, for the application we have mentioned, only a good lower bound is needed for the number $I_{a,\vartheta}(N, H)$ of i with $0 \leq i \leq p/H - 1$ such that $T_{a,\vartheta}(N, iH, H) > 0$. Of course, trivially,

$$\min\{p/H, N\} \geq I_{a,\vartheta}(N, H) \geq \max\{N/H - 1, 0\}. \tag{15.4}$$

We shall show how bounds for the average value $R_{a,\vartheta}(N, H)$, like the two above, allow us to get better lower bounds for $I_{a,\vartheta}(N, H)$.

First of all, we mention the bound

$$I_{a,\vartheta}(N, H) = \frac{p}{H}\left(1 + O\left(Hp^{-1} + p^2 R_{a,\vartheta}(N, \lfloor H/2 \rfloor)N^{-2}H^{-2}\right)\right)$$

communicated to us by Hugh Montgomery [58].

Indeed, since $T_{a,\vartheta}(N, iN, H) = 0$ for at least $p/H - 1 - I_{a,\vartheta}(N, H)$ values of i with $0 \leq i \leq p/H - 1$, then evidently $T_{a,\vartheta}(N, M, \lfloor H/2 \rfloor) = 0$ for at least $0.5 \left(p/H - 1 - I_{a,\vartheta}(N, H) \right) H$ values of M. Hence

$$0.5 \left(p/H - 1 - I_{a,\vartheta}(N, H) \right) H (NH/p)^2 \leq p R_{a,\vartheta}(N, \lfloor H/2 \rfloor).$$

Taking into account the trivial upper bound $I_{a,\vartheta}(N, H) \leq p/H$, we get the desired result.

Now we can apply any of the bounds above of $R_{a,\vartheta}(N, H)$. For example, substituting the bounds (15.2) gives the asymptotic result $I_{a,\vartheta}(N, H) \sim p/H$ for $N^2 H \geq p^{2+\varepsilon}$. Unfortunately, when $NH \sim p$ even (15.3) does not give anything.

We shall now obtain another lower bound for $I_{a,\vartheta}(N, H)$ in terms of $R_{a,\vartheta}(N, H)$ which is non-trivial in a wider range of parameters (including $NH \sim p$).

Lemma 15.1. *The bound*

$$I_{a,\vartheta}(N, H) + 1 \gg \min \left\{ \frac{p}{H}, \frac{N^2 H}{p R_{a,\vartheta}(N, H)} \right\}$$

holds.

Proof For simplicity, we suppose that H is even, but it is easy to see that this is not essential.

Suppose that $H > p/10$ and $I_{a,\vartheta}(N, H) < p/5H - 1$, otherwise there is nothing to prove.

Denote by $\mathcal{I}_{a,\vartheta}(N, H, t)$ the set of i with $0 \leq i \leq p/H$ such that $T_{a,\vartheta}(N, iH, H) = t$. Because we have included one extra interval (corresponding to $i = \lfloor p/H \rfloor$) in our considerations, we see that

$$I_{a,\vartheta}(N, H) \leq \sum_{t=1}^{N} |\mathcal{I}_{a,\vartheta}(N, H, t)| \leq I_{a,\vartheta}(N, H) + 1.$$

We also have

$$\sum_{t=1}^{N} t |\mathcal{I}_{a,\vartheta}(N, H, t)| = N.$$

Let us consider the sums

$$W_1 = \sum_{t > 4NH/p} t |\mathcal{I}_{a,\vartheta}(N, H, t)|, \qquad W_2 = \sum_{t > 4NH/p} t^2 |\mathcal{I}_{a,\vartheta}(N, H, t)|.$$

We have

$$W_1 \geq N - \frac{4NH}{p} \sum_{t \leq 4NH/p} |\mathcal{I}_{a,\vartheta}(N, H, t)| \geq N - \frac{4NH\left(I_{a,\vartheta}(N, H) + 1\right)}{p}.$$

Thus from our assumption, we derive

$$W_1 > N/5. \tag{15.5}$$

If $T_{a,\vartheta}(N, iH, H) = t$, then either

$$T_{a,\vartheta}(N, iH, H/2) \geq t/2$$

or

$$T_{a,\vartheta}(N, iH + H/2, H/2) \geq t/2.$$

In the first case, we set $\Delta_i(t) = [iH - H/2, iH - 1]$, in the second case, $\Delta_i(t) = [iH, iH + H/2 - 1]$. Note that each residue modulo p belongs to at most two of the intervals $\Delta_i(t)$, $i = 0, \ldots, \lfloor p/H \rfloor$. In fact, only the intervals $\Delta_0(t)$ and $\Delta_{\lfloor p/H \rfloor}(t)$ can overlap.

Further, for $T_{a,\vartheta}(N, iH, H) = t > 4NH/p$ and $M \in \Delta_i(t)$, we have

$$T_{a,\vartheta}(N, M, H) - NH/p \geq t/2 - NH/p > t/4.$$

Obviously,

$$\sum_{i \in \mathcal{I}_{a,\vartheta}(N,H,t)} |\Delta_i(t)| = 0.5 |\mathcal{I}_{a,\vartheta}(N, H, t)| H.$$

Summing over all $t > 4NH/p$ and $M \in \Delta_i(t)$, $0 \leq i \leq p/H$, we obtain

$$
\begin{aligned}
R_{a,\vartheta}(N, H) &\geq \frac{1}{2p} \sum_{t > 4NH/p} \sum_{i \in \mathcal{I}_{a,\vartheta}(N,H,t)} \sum_{M \in \Delta_i(t)} t^2/16 \\
&\geq \frac{H}{64p} \sum_{t > 4NH/p} t^2 |\mathcal{I}_{a,\vartheta}(N, H, t)| = \frac{H}{64p} W_2.
\end{aligned}
$$

From this inequality and from (15.5), we derive

$$I_{a,\vartheta}(N, H) + 1 \geq \sum_{t > 4NH/p} |\mathcal{I}_{a,\vartheta}(N, H, t)| \geq \frac{W_1^2}{W_2} \gg \frac{N^2 H}{p R_{a,\vartheta}(N, H)},$$

and the statement follows. $\qquad\qquad\qquad\qquad\qquad\qquad\qquad\qquad\qquad\square$

From the above mentioned upper bounds of $R_{a,\vartheta}(N, H)$, we obtain

$$
I_{a,\vartheta}(N, H) + 1
$$
$$
\gg \begin{cases} N, & \text{if } N \geq p^{5/7+\varepsilon}; \ NH \sim p \\ \min\{N^2 p^{-1+\varepsilon}, pH^{-1}\}, & \text{if } 1 \leq N \leq p - 1, \ 1 \leq H \leq p. \end{cases}
$$

Note that the first bound is the best possible and improves the trivial lower bound for all allowed values of N and H. The second bound does the same if $NH > p^{1+\varepsilon}$.

As we have discussed, for applications of these bounds to the mentioned sorting algorithm, we have to consider the case when $NH \sim p$. So the second bound cannot be useful but the first one allows us to get a partial improvement of the lower bound of Theorem 4 of [36] (see also Theorem 12.1 of [90]).

Now we propose quite a different method to deal with $I_{a,\vartheta}(N, H)$.

Lemma 15.2. *For any integer $k \geq 2$, the bound*

$$I_{a,\vartheta}(N, H) + 1 \gg Np^{-\varepsilon} \frac{\min\{p^{1/k}H^{-1}, 1\}}{\min\{N^{1/k}, N^{1/k}Hp^{-1/k} + p^{1/2k}\}}$$

holds.

Proof It is easy to see that there exists at least one v, $1 \leq |v| < N$, such that the congruence

$$a(\vartheta^{x+v} - \vartheta^x) \equiv y \pmod{p},$$
$$\max\{1, -v\} \leq x \leq \min\{N, N-v\}, \qquad 1 \leq y < H,$$

has at least

$$J \gg \frac{N}{I_{a,\vartheta}(N, H)} + 1$$

solutions.

Let $h \equiv a(\vartheta^v - 1) \pmod{p}$, $1 \leq h \leq p$, then there exists a subset $S \subseteq \{1, \ldots, H\}$ containing J elements such that any element of S is congruent to $\vartheta^x h \pmod{p}$ for some x, $1 \leq x \leq N$. Since any t has $O(p^\varepsilon)$ representations in the form $s_1 \ldots s_k$, $s_1, \ldots, s_k \in \mathbb{Z}$, the set

$$T = \{s_1 \ldots s_k \ : \ s_1, \ldots, s_k \in S\}$$

contains at least $K \gg J^k p^{-\varepsilon}$ elements of the interval $[1, H^k]$. Therefore, T contains at least

$$L \gg K \min\{1, p/H^k\} \gg N^k I_{a,\vartheta}(N, H)^{-k} p^{-\varepsilon} \min\{1, p/H^k\}$$

different residues modulo p. On the other hand, any element of T is congruent to $\vartheta^x h^k$ for some x with $1 \leq x \leq kN$, thus $L \leq kN$. Also, evidently $L \leq T_{a,\vartheta}(N, H^k, 1)$, hence

$$L \ll \min\{N, NH^k/p + p^{1/2+\varepsilon}\}$$

and we have the required estimate for $I_{a,\vartheta}(N, H)$. $\qquad \square$

The above mentioned upper bounds on $R_{a,\vartheta}(N, H)$ together with Lemmas 15.1 and 15.2 imply the following improvement of the trivial lower bound (15.4).

Theorem 15.3 *For sufficiently large p, the bounds*

$$I_\vartheta(N, \lfloor p/N \rfloor) \gg \begin{cases} N, & \text{if } N \geq p^{5/7+\varepsilon}; \\ N^{11/4} p^{-5/4-\varepsilon}, & \text{if } p^{4/7} \leq N < p^{5/7+\varepsilon}; \\ N p^{-1/4-\varepsilon}, & \text{if } p^{1/2} \leq N < p^{4/7}; \\ N^{3/2} p^{-1/2-\varepsilon}, & \text{if } p^{1/3+\varepsilon} \leq N < p^{1/2}; \end{cases}$$

hold.

To get this result, we combine Lemma 15.1 and the bound (15.3) (for the first and the second estimates) and use Lemma 15.2 with $k = 2$ (for the third and fourth estimates).

Question 15.4. *Can these three different approaches shown above be combined in one bound which supersedes the previous ones?*

Although we have improved the trivial lower bound (15.4) for $N \geq p^{1/3+\varepsilon}$, unfortunately only results concerning values $N \geq p^{3/5}$ are applicable to the QuickSort algorithm.

Indeed, it is shown in [36, 90] that the lower bound of the expected running time of the 'pseudo-randomized' QuickSort algorithm is of the form

$$\Omega\left(I_{a,\vartheta}(N, \lfloor p/N \rfloor) N^2 p^{-1} + N \ln N\right).$$

So, if we wish to obtain the running time of order $N \ln N$ (that is, the running time which is expected with the QuickSort algorithm using ideal random numbers), we have to use a generator modulo p of order at least $N^{5/3}$ while the trivial estimate (15.4) gives $N^{3/2}$ and the hypothetical bound $I_{a,\vartheta}(N, \lfloor p/N \rfloor) \gg N$ would imply $p \gg N^2$ (everywhere up to some slower growing terms). Below we show that we really need p of this order for almost all initial values $a = 1, \ldots, p - 1$ but firstly we consider the situation when the primitive root ϑ is taken at random. The question of estimating the average value

$$\rho_a(N, H) = \frac{1}{\varphi(p-1)} \sum_\vartheta R_{a,\vartheta}(N, H)$$

over all $\varphi(p - 1)$ primitive roots ϑ is mentioned in [57] where a hope is expressed that for $\rho_a(N, H)$, a good estimate or even an asymptotic formula can be obtained for essentially smaller values of N than is needed for individual values of $R_{a,\vartheta}(N, H)$. It turns out that it can really be done for the values of $N \leq p^{1/4-\varepsilon}$.

We need the following lemma which is formulated in a much more general form than is necessary for our purpose.

Lemma 15.5. *For any non-constant polynomial* $f \in \mathbb{F}_p^*[x]$ *of degree* $n > 0$, *the bound*

$$\left| \sum_\vartheta \mathbf{e}\left(f(\vartheta)/p\right) \right| \leq 2^{\omega(p-1)} n p^{1/2}$$

holds, where the sum is taken over all $\varphi(p-1)$ *primitive roots* ϑ *modulo* p.

Proof Let us fix a primitive root g modulo p. Then

$$\sum_\vartheta \mathbf{e}\left(f(\vartheta)/p\right) = \sum_{\substack{t=1 \\ \gcd(t, p-1)=1}} \mathbf{e}\left(f(g^t)/p\right).$$

Using the following known property:

$$\sum_{d|m} \mu(d) = \begin{cases} 1, & \text{if } m = 1; \\ 0, & \text{if } m \geq 2; \end{cases}$$

of the Möbius function $\mu(k)$ we obtain

$$\sum_\vartheta \mathbf{e}\left(f(\vartheta)/p\right) = \sum_{m=1}^{p-1} \sum_{d|m} \mu(d) \sum_{\substack{t=1 \\ \gcd(t, p-1)=m}}^{p-1} \mathbf{e}\left(f(g^t)/p\right)$$

$$= \sum_{d|p-1} \mu(d) \sum_{\substack{m=1, \\ d|m}}^{p-1} \sum_{\substack{t=1 \\ \gcd(t, p-1)=m}}^{p-1} \mathbf{e}\left(f(g^t)/p\right) = \sum_{d|p-1} \mu(d) \sum_{\substack{t=1 \\ d|t}}^{p-1} \mathbf{e}\left(f(g^t)/p\right)$$

$$= \sum_{d|p-1} \frac{\mu(d)}{d} \sum_{t=1}^{p-1} \mathbf{e}\left(f(g^{dt})/p\right) = \sum_{d|p-1} \frac{\mu(d)}{d} \sum_{x=1}^{p-1} \mathbf{e}\left(f(x^d)/p\right).$$

The Weil bound (see Chapter 5 of [52]) entails that each inner sum does not exceed $dnp^{1/2}$, therefore,

$$\sum_{\substack{t=1 \\ \gcd(t, p-1)=1}} \mathbf{e}\left(f(g^t)/p\right) \leq \deg f p^{1/2} \sum_{d|p-1} |\mu(d)|.$$

The last sum is the number of square free divisors of $p - 1$ which equals $2^{\omega(p-1)}$.

                                                            □

Theorem 15.6 *The bound*

$$\rho_a(N, H) = NHp^{-1} + O(NH^2p^{-2} + N^3Hp^{-3/2+\varepsilon})$$

holds.

Proof Applying standard arguments, we get

$$
\begin{aligned}
T_{a,\vartheta}(N, M, H) - \frac{NH}{p} &= \frac{1}{p}\sum_{\alpha=1}^{p-1}\sum_{x=1}^{N}\mathbf{e}(a\alpha\vartheta^x/p)\sum_{u=M}^{M+H-1}\mathbf{e}(-\alpha u/p) \\
&= \frac{1}{p}\sum_{\alpha=1}^{p-1}\mathbf{e}(-\alpha M/p) \\
&\quad \times \sum_{x=1}^{N}\mathbf{e}(a\alpha\vartheta^x/p)\sum_{u=0}^{H-1}\mathbf{e}(-\alpha u/p).
\end{aligned}
$$

Therefore,

$$
\begin{aligned}
R_{a,\vartheta}(N, H) &= \frac{1}{p^3}\sum_{\alpha,\beta=1}^{p-1}\sum_{x,y=1}^{N}\mathbf{e}\left(a(\alpha\vartheta^x - \beta\vartheta^y)/p\right) \\
&\quad \times \sum_{u,v=0}^{H-1}\mathbf{e}\left((\beta v - \alpha u)/p\right)\sum_{M=0}^{p-1}\mathbf{e}\left((\beta - \alpha)M/p\right).
\end{aligned}
$$

The inner sum is 0 if $\alpha \not\equiv \beta \pmod{p}$ and equals p otherwise. Hence

$$R_{a,\vartheta}(N, H) = \frac{1}{p^2}\sum_{\alpha=1}^{p-1}\left|\sum_{x=1}^{N}\mathbf{e}(a\alpha\vartheta^x/p)\right|^2\left|\sum_{u=0}^{H-1}\mathbf{e}(\alpha u/p)\right|^2. \qquad (15.6)$$

From Lemma 15.5, we find

$$
\begin{aligned}
\sum_{\vartheta}\left|\sum_{x=1}^{N}\mathbf{e}(a\alpha\vartheta^x/p)\right|^2 &= \sum_{x,y=1}^{N}\sum_{\vartheta}\mathbf{e}\left(a\alpha(\vartheta^x - \vartheta^y)/p\right) \\
&= N\varphi(p-1) + \sum_{\substack{x,y=1\\x\neq y}}^{N}\sum_{\vartheta}e\left(a\alpha(\vartheta^x - \vartheta^y)/p\right) \\
&= N\varphi(p-1) + O(2^{\omega(p-1)}N^3p^{1/2}).
\end{aligned}
$$

Combining this inequality and (15.6), we obtain

$$
\begin{aligned}
\rho_a(N, H) &= \frac{1}{\varphi(p-1)p^2} \sum_{\alpha=1}^{p-1} \sum_{\vartheta} \left| \sum_{x=1}^{N} \mathbf{e}(a\alpha\vartheta^x/p) \right|^2 \left| \sum_{u=0}^{H-1} \mathbf{e}(\alpha u/p) \right|^2 \\
&\leq \frac{N\varphi(p-1) + O(2^{\omega(p-1)}N^3 p^{1/2})}{\varphi(p-1)p^2} \sum_{\alpha=1}^{p-1} \left| \sum_{u=0}^{H-1} \mathbf{e}(\alpha u/p) \right|^2 \\
&= \frac{N\varphi(p-1) + O(2^{\omega(p-1)}N^3 p^{1/2})}{\varphi(p-1)p^2} (pH - H^2).
\end{aligned}
$$

Thus, after simple evaluations, taking into account that $2^{\omega(p-1)}p/\varphi(p-1) = O(p^\varepsilon)$ we obtain the assertion. $\qquad\square$

Therefore if $NH \sim p$, Theorem 15.6 implies $\rho_a(N, H) \sim 1$ for $p^\varepsilon \leq N \leq p^{1/4-\varepsilon}$ (and of course for $N \geq p^{5/7+\varepsilon}$ because of Theorem 15.3).

From Lemma 15.1 and Theorem 15.6, we see that

$$
I_{a,\vartheta}(N, H) \gg \min\{N, \ p/H, \ p^{1/2-\varepsilon}N^{-1}\}
$$

for all but $\delta\varphi(p-1)$ primitive roots ϑ, for any fixed $\delta > 0$. Since $I_{a,\vartheta}(N, H)$ is an increasing function of N for fixed H, we find that

$$
I_{a,\vartheta}(N, \lfloor p/N \rfloor) \gg \begin{cases} N, & \text{if } N < p^{1/4-\varepsilon}; \\ p^{1/4-\varepsilon} & \text{if } N \geq p^{1/4-\varepsilon} \end{cases}
$$

for all but $\delta\varphi(p-1)$ primitive roots ϑ, for any fixed $\delta > 0$. It improves Theorem 15.3 for any $N \leq p^{1/2-\varepsilon}$. Unfortunately, it does not include any new results for the QuickSort algorithm.

Of course, it would be very interesting to fill the gap between $p^{1/4-\varepsilon}$ and $p^{5/7+\varepsilon}$.

Question 15.7. *Prove the asymptotic formula $\rho_a(N, H) \sim NH/p$ for any $N \geq p^\varepsilon$.*

Finally, we observe that similar considerations allow us to evaluate explicitly the average values of $R_{a,\vartheta}(N, H)$ over $a = 1, \ldots, p-1$,

$$
\frac{1}{p} \sum_{a=1}^{p-1} R_{a,\vartheta}(N, H) = NH \frac{(p-N)(p-H)}{p^3}.
$$

Indeed, to get it, we use the identity

$$
\sum_{\alpha=1}^{p-1} \left| \sum_{x=1}^{N} \mathbf{e}(\alpha\vartheta^x) \right|^2 = N(p-N)
$$

instead of Lemma 15.5 in the proof of Theorem 15.6. Therefore,

$$I_{a,\vartheta}(N, \lfloor p/N \rfloor) \geq N p^{-\varepsilon}$$

for all but possibly at most $O(p^{1-\varepsilon})$ values $a = 1, \ldots, p - 1$. Thus for all but possibly at most $O(p^{1-\varepsilon})$ values $a = 1, \ldots, p - 1$ the complexity of the corresponding QuickSort algorithm is $\Omega(N^3 p^{-1-\varepsilon} + N \ln N)$ and to obtain the desired order $N \ln N$, we necessarily should select p of order at least $N^{2-\varepsilon}$.

Part seven

Applications to Coding Theory and Combinatorics

16

Difference Sets in $V_{\mathfrak{p}}$

In this chapter we consider some multiplicative properties of difference sets, that is, sets of the form

$$D = \{a - b \ : \ a, b \in S\}$$

where S is a certain set.

In [49], Hendrik Lenstra considers the largest integer M such that there exists a set $\Omega = \{\omega_1, \ldots, \omega_M\} \subseteq \mathbb{Z}_{\mathbb{K}}$ with

$$\omega_i - \omega_j \in U, \qquad 1 \le i, j \le M, \ i \ne j,$$

where U is the unit group of $\mathbb{K}$. Denote this largest M by $M(\mathbb{K})$. If $M(\mathbb{K})$ is large enough, then $\mathbb{K}$ is Euclidean. Hendrik Lenstra also obtains some upper and lower bounds for $M(\mathbb{K})$.

The same functions $M(\mathbb{K}, V)$ can be defined with respect to any finitely generated multiplicative group V of $\mathbb{K}$. But in this more general setting, it is not quite clear how to apply the method of [49] (see also [48, 65, 70]) to obtain upper and lower bounds for $M(\mathbb{K}, V)$.

It is interesting to note that these functions $M(\mathbb{K}, V)$ are related to a question about the cycles in polynomial mappings over algebraic number field (see Chapter 12 of [66]).

Here we consider an analogue of these functions modulo an integer ideal q.

Define $M(\mathbb{K}, V, \mathfrak{q})$ as the largest integer M such that there exists a set

$$\Omega = \{\omega_1, \ldots, \omega_M\} \subseteq \Lambda_{\mathfrak{q}}$$

with

$$\omega_i - \omega_j \in V_{\mathfrak{q}}, \qquad 1 \le i, j \le M, \ i \ne j. \tag{16.1}$$

In particular, it follows from the definition that if $-1 \notin V_{\mathfrak{q}}$, then $M(\mathbb{K}, V, \mathfrak{q}) = 1$.

143

We obtain an upper bound for $M(\mathbb{K}, V, \mathfrak{p})$ for prime ideals $\mathfrak{p}$. Using this bound and the trivial inequality $M(\mathbb{K}, V) \leq M(\mathbb{K}, V, \mathfrak{p})$, for an appropriate prime ideal $\mathfrak{p}$, one may get an upper bound for $M(\mathbb{K}, V)$.

Theorem 16.1 *Let* $\mathbb{K}$ *have at least one principal unit. If* $V_{\mathfrak{p}} = \Lambda_{\mathfrak{p}}^{*}$, *then* $M(\mathbb{K}, V, \mathfrak{p}) = \mathrm{Nm}(\mathfrak{p})$. *Otherwise*

$$M(\mathbb{K}, V, \mathfrak{p}) \leq \begin{cases} 2\,\mathrm{Nm}(\mathfrak{p})^{1/2} + 2, & \text{for any } \mathfrak{p}, \\ 4|V_{\mathfrak{p}}|^{2/3} + 2, & \text{if } \mathfrak{p} \text{ is of first degree.} \end{cases}$$

Proof The statement is trivial if $V_{\mathfrak{p}} = \Lambda_{\mathfrak{p}}^{*}$. In the other case, we have $|V_{\mathfrak{p}}| \leq |\mathrm{Nm}(\mathfrak{p})|/2$, because $V_{\mathfrak{p}}$ is a subgroup of $\Lambda_{\mathfrak{p}}^{*}$. Furthermore, suppose that for some set

$$\Omega = \{\omega_1, \dots, \omega_M\} \subseteq \Lambda_{\mathfrak{p}},$$

we have (16.1). Then the congruence

$$\omega_i - \omega_j \equiv u \pmod{\mathfrak{p}}, \qquad u \in V_{\mathfrak{p}}, \ 1 \leq i, j \leq M,$$

has exactly $N = M(M - 1)$ solutions.
By using the same considerations as in the proof of Lemma 9.2, one sees that

$$\left| N - M^2 |V_{\mathfrak{p}}| / \mathrm{Nm}(\mathfrak{p}) \right| \leq M\,\mathrm{Nm}(\mathfrak{p})^{1/2}.$$

On the other hand,

$$N - M^2 |V_{\mathfrak{p}}| / \mathrm{Nm}(\mathfrak{p}) \geq M(M - 1) - M^2/2 = M(M/2 - 1)$$

and the first bound follows.
To derive the second bound, we mention that for $a = \omega_2 - \omega_1$ there are at least $M(\mathbb{K}, V, \mathfrak{p}) - 2$ representations $a \equiv u + v \pmod{p}$ (where $p = \mathrm{Nm}(\mathfrak{p})$) with $u = \omega_2 - \omega_j \in V_{\mathfrak{p}}, v = \omega_j - \omega_1 \in V_{\mathfrak{p}}, j = 3, \dots, M(\mathbb{K}, V, \mathfrak{p})$. Thus if

$$|V_{\mathfrak{p}}| \leq \frac{p - 1}{(p - 1)^{1/4} + 1},$$

then from (3.8) we obtain $M(\mathbb{K}, V, \mathfrak{p}) \leq 4|V_{\mathfrak{p}}|^{2/3} + 2$ (for larger $|V_{\mathfrak{p}}|$ the first bound is stronger). □

It is also evident that the same considerations which we have used for prime ideals of first degree imply $M(\mathbb{K}, V, \mathfrak{q}) \leq |V_{\mathfrak{q}}| + 1$ for any integer ideal $\mathfrak{q}$.

Theorem 16.1 is based on the bound (3.8) on the cardinality of the set

$$S_a = \{x : x \in V_{\mathfrak{p}}, a - x \in V_{\mathfrak{p}}\}.$$

For the problem considered in this chapter, it would be useful to generalize this result.

Question 16.2. *Obtain an upper bound on the cardinality of the set*

$$S_{ab} = \{x \;:\; x \in V_{\mathfrak{p}}, \; a - x \in V_{\mathfrak{p}}, \; b - x \in V_{\mathfrak{p}}\},$$

where a, b are distinct elements from $\Lambda_{\mathfrak{p}}$, which is stronger than the bound which follows from (3.8).

In the case where $\mathbb{K} = \mathbb{Q}$, $\mathfrak{q} = q$ is a rational integer and V_q is the set of quadratic residues modulo q, this function has been considered in [14]. In particular, if $\mathbb{K} = \mathbb{Q}$ and $\mathfrak{p} = p$ is a rational prime, a lower bound of order $\ln p$ holds (see Theorem 3 of [14]). For groups of large size $|V_{\mathfrak{p}}|$ (of order $\mathrm{Nm}(\mathfrak{p})$, say) it can be extended to the case of $M(\mathbb{K}, V, \mathfrak{p})$ as well. However, for smaller $|V_{\mathfrak{p}}|$ (of order $\mathrm{Nm}(\mathfrak{p})^{1-\varepsilon}$, say) the approach of that paper does not work. We assume that actually $M(\mathbb{K}, V, \mathfrak{p})$ is bounded for such 'small' groups.

Question 16.3. *Is it true that for any $\varepsilon > 0$, there is a constant $C(\varepsilon)$ such that $M(\mathbb{K}, V, \mathfrak{p}) \le C(\varepsilon)$ for $|V_{\mathfrak{p}}| < \mathrm{Nm}(\mathfrak{p})^{1-\varepsilon}$?*

At the moment we can only prove that for $\mathbb{K} = \mathbb{Q}$, there are infinitely many primes p and groups V with $|V_p| > cp^{1/3}$ where $c > 0$ is an absolute constant and such that $M(\mathbb{Q}, V, p) = 2$.

Indeed, it is similar to the proof of Theorems 5.5 and 14.7. Let $r \ge 5$ be a prime and let W_p be a subgroup of $\mathbb{F}_p^*$ of cardinality $|W_p| = r$ and let $V_p = W_p \cup -W_p$. Evidently, $-1 \notin W_p$ thus $|V_p| = 2r$. If $M(\mathbb{K}, V, p) > 2$ (where V is obtained from V_p by lifting), then $v_1 + v_2 + v_3 = 0$ for some $v_1, v_2, v_3 \in V_p$ and hence $u + w + 1 = 0$ for some $u, w \in V_{\mathfrak{p}}$. At least one of u and w, say w, is not equal to ± 1 and thus is of multiplicative order r or $2r$. Therefore $u = \pm w^k$ with some integer k, $0 \le k \le 2r - 1$. So, we have $\pm w^k + w + 1 \equiv 0 \pmod{p}$ and $w^{2r} - 1 \equiv 0 \pmod{p}$. Therefore p is a factor of the resultant $R_{r,k}$ of polynomials $x^{2r} - 1$ and $1 + x \pm x^k$ with some $0 \le k \le 2r - 1$. First of all, we show that $R_{r,k} \ne 0$. Indeed, if $R_{r,k} = 0$, then for some $2r$th root of unity ρ we have $|1 + \rho| = |\rho^k| = 1$. Simple geometric arguments show that the only possibility is $\rho = \mathbf{e}(\pm 1/3)$ which is impossible since $r \ge 5$. On the other hand, we evidently have $\ln |R_{r,k}| = O(r)$. Therefore, the number of primes p dividing $R_{r,k}$ is $O(r/\ln r)$, for every $k = 0, \ldots, 2r - 1$. Thus the number of acceptable primes p (that is, primes with $M(\mathbb{K}, V, \mathfrak{p}) > 2$) does not exceed $Cr^2/\ln r$ with some absolute constant $C > 0$. It follows from Theorem 2.1 of [1] that there are infinitely many primes r for which the number of primes p such that $p \equiv 1 \pmod{r}$ and $p < 4Cr^3$ is greater than $Cr^2/\ln r$. Hence for at least one $p < 4Cr^3$ we have $M(\mathbb{Q}, V, p) = 2$.

We also remark that if $|V_p|$ is divisible by 6, then $M(\mathbb{K}, V, p) \geq 3$. Indeed, if $v \in V_p$ is of multiplicative order 6, then $1 + v^2 \equiv v \pmod{\mathfrak{p}}$, thus $\{0, 1, -v^2\}$ is such a difference set.

On the other hand, for the case $\mathbb{K} = \mathbb{Q}$ one can state that if $|V_p| \geq 4$ is not divisible by 6 and

$$\mathrm{Nm}(\mathfrak{p}) > 6^{|V_p|/4},$$

then $M(\mathbb{K}, V, \mathfrak{p}) \leq 2$.

To prove this inequality, we denote $p = \mathrm{Nm}(\mathfrak{p})$, $t = |V_p|$. As we have remarked, if $-1 \notin V_p$, then $M(\mathbb{K}, V, \mathfrak{p}) = 1$. Therefore we can assume that t is even. Let g be an element of V_p of order t, thus $g^{t/2} + 1 \equiv 0 \pmod{p}$. Assume that $M(\mathbb{K}, V, \mathfrak{p}) \geq 3$. This gives the solvability of the congruence $g^A + g^B \equiv g^C \pmod{p}$ for some integers A, B, C. Taking residues of A, B, C modulo $t/2$, we can easily deduce the existence of non-negative integers a, b such that $a < t/2$, $b < t/2$ and $1 \pm g^a \pm g^b \equiv 0 \pmod{p}$. If there are two equal numbers among $0, a, b$, then $2 \in V_p$ and $2^t - 1 \equiv 0 \pmod{p}$. Therefore either $2^{t/2} - 1 \equiv 0 \pmod{p}$ or $2^{t/2} + 1 \equiv 0 \pmod{p}$ which is not possible because $p > 6^{t/4} \geq 2^{t/2} + 1$. Now we assume, that $0, a, b$ are distinct. The resultant of the polynomials $P(z) = z^{t/2} + 1$ and $Q(z) = 1 \pm z^a \pm z^b$ (both of degree at most $t/2$) is divisible by p. On the other hand, this resultant is not equal to zero, because $t \not\equiv 0 \pmod{6}$, and does not exceed $6^{t/4}$. This proves the statement.

By similar but technically more complicated reasons, one can show that $M(\mathbb{K}, V, \mathfrak{p}) \leq 3$ for $p > C^t$ where $C > 1$ is a constant (even if t is divisible by 6).

It is interesting to note that a question about the size of $M(\mathbb{K}, V, \mathfrak{p})$ arises in algebraic combinatorics as well [53] (at least when V_p is the group of quadratic residues). In particular, Lemma 10.4.16 of [53] provides a construction showing that the square root bound of Theorem 16.1 is actually the best possible for general $\mathfrak{p}$ and V. This construction is so elegant that we present it here.

Let $\mathfrak{p}$ be a prime ideal of even degree, $\mathrm{Nm}(\mathfrak{p}) = p^{2m} > 4$. Then each element of $\mathbb{F}_{p^m}^*$ is a quadratic residue of $\Lambda_\mathfrak{p} \simeq \mathbb{F}_{p^{2m}}$. Thus, the set $\Omega = \mathbb{F}_{p^m}$ shows that $M(\mathbb{K}, Q, \mathfrak{p}) \geq \mathrm{Nm}(\mathfrak{p})^{1/2}$ where Q is a group generated by quadratic residues modulo $\mathfrak{p}$.

Now, for a set

$$\Omega = \{\omega_1, \ldots, \omega_M\} \subseteq \mathbb{Z}_{\mathbb{K}},$$

we consider the directed graphs $\mathcal{G}(\Omega, V)$ and $\mathcal{G}(\Omega, V, \mathfrak{q})$ which have M

vertices labeled by $\omega_1, \dots, \omega_M$, such that vertices ω_i and ω_j are connected if and only if $\omega_i - \omega_j \in V$ and $\omega_i - \omega_j \in V_\mathfrak{q}$, respectively.

Obviously, $M(\mathbb{K}, V)$ and $M(\mathbb{K}, V, \mathfrak{p})$ are just the maximal values of M such that there exist sets Ω with $|\Omega| = M$ for which the graphs $\mathcal{G}(\Omega, V)$ and $\mathcal{G}(\Omega, V, \mathfrak{p})$ are each M-vertex complete graphs, respectively. Various other problems concerning graphs $\mathcal{G}(\Omega, V)$ are dealt with in [29, 30].

Question 16.4. *Obtain analogues of results of papers [29, 30] for graphs $\mathcal{G}(\Omega, V, \mathfrak{p})$.*

Then it is easy to see that the question about the diameter of the graph $\mathcal{G}(\Lambda_\mathfrak{p}, V, \mathfrak{p})$ can be reformulated as a variant of the Waring problem for $\mathbb{F}_q$, where $q = \mathrm{Nm}(\mathfrak{p})$, for the power $s = (\mathrm{Nm}(\mathfrak{p}) - 1)/|V_\mathfrak{p}|$. For the first time, it was apparently remarked in [18]. A survey of known results can be found in [41] and in Section 5.1 of [84]. The majority of these results deal with the particular case of a prime field (that is, when $\mathfrak{p}$ is a prime ideal of first degree).

Question 16.5. *Study graphs $\mathcal{G}(\Omega, V, \mathfrak{q})$ for arbitrary integer ideals $\mathfrak{q}$.*

17

Dimension of BCH Codes

Here we show how our methods can be applied to the study of one of the most attractive and widely usable codes, namely BCH codes.

We start this chapter by recalling some definitions [5, 54].

Let q be a power of a prime and let n be an integer with $\gcd(n, q) = 1$ such that q is of multiplicative order t modulo n.

Then $\mathbb{F}_{q^t}$ contains an element $\alpha \in \mathbb{F}_{q^t}^*$ of multiplicative order n. Let l be an integer. To construct a BCH code with *constructive distance* Δ we consider the polynomial g over $\mathbb{F}_q$ of the smallest degree such that

$$g\left(\alpha^{l+y}\right) = 0, \qquad y = 1, \ldots, \Delta - 1,$$

and consider the cyclic code of length n with g as the generator polynomial, that is the linear space of dimension $k = n - \deg g$ of n-dimensional vectors $(a_0, \ldots, a_{n-1}) \in \mathbb{F}_q^n$ such that

$$a_0 + a_1 Z + \cdots + a_{n-1} Z^{n-1} \equiv 0 \pmod{g(Z)}.$$

Generally, for every code, there are three parameters of interest: the length, the minimal distance and the dimension. For a BCH code, the length n is given, the minimal distance d is at least the constructive distance Δ (and this bound is known to be tight in many cases [5, 54]). The question about the dimension is more interesting. Of course, $t \leq \deg g \leq Dt$, thus the dimension $n - t \geq k \geq n - (\Delta - 1)t$. To get something stronger one should study the structure of the roots of g in more detail.

First of all, we make an observation that all roots of g are powers of α because trivially

$$g(Z) \mid \prod_{y=1}^{\Delta-1} \prod_{x=1}^{t} \left(Z - \alpha^{(l+y)q^x}\right).$$

148

We also remark that α^j is a root of g if and only if

$$jq^x \equiv l + y \pmod{n},$$

for some $x = 1, \ldots, t$ and $y = 1, \ldots, \Delta - 1$.

Let us denote by $J(q, n, \Delta)$ the maximal dimension of q-ary generalized BCH codes of length n and of designed distance Δ taken over all $l = 0, \ldots, n-1$. Then we see that for some integer l, $J(q, n, \Delta)$ equals the number of $j = 0, 1, \ldots, n - 1$ for which the congruence

$$jq^x \equiv l + y \pmod{n}, \qquad x = 1, \ldots, t, \quad y = 1, \ldots, \Delta, \qquad (17.1)$$

is not solvable.

Thus the original question has been reduced to a question about the distribution of values of an exponential function to which our technique can be applied.

The bound

$$J(q, n, \Delta) \le \frac{3n^3}{(\Delta - 1)^2 t^{1/2}}$$

is stated in [80]. For a wide range of parameters, it has been improved in [84] to

$$J(q, n, \Delta) \le \frac{24n^5}{(\Delta - 1)^4 t}.$$

Both these results are based on some bounds of exponential sums with exponential functions. Here we obtain a further improvement. The method we use is a modification of that used in the proof of Theorem 13.1.

For a divisor d of n denote by t_d the multiplicative order of q modulo d (thus $t = t_n$). The following statement is Lemma 3 of [80].

Lemma 17.1. *For any $d \mid n$, the bound $t_{n/d} \ge t/d$ holds.*

Lemma 17.2. *For any integers a, b, the congruence*

$$aq^x \equiv bq^y \pmod{n}, \qquad 1 \le x, y \le t$$

is solvable only when $\gcd(a, n) = \gcd(b, n) = d$, and in this case for the number of solutions $N(a, b)$, the bound

$$N(a, b) \le td$$

holds.

Proof As $\gcd(q, n) = 1$, the condition on a and b is evident. Also it is evident that for any fixed x, there are at most $t/t_{n/d}$ possible values for y, hence $N(a, b) \le t^2/t_{n/d} \le td$ because of Lemma 17.1. $\square$

We define the sums

$$T(a, h) = \sum_{u=1}^{h} e(au/n), \qquad W_d(h) = \sum_{\gcd(a,n)=d} |T(a, h)|^2,$$

where d is a divisor of n, $d \mid n$.

Lemma 17.3. *For any $d \mid n$ with $d < n$, the bound*

$$W_d(h) \leq nh/d$$

holds.

Proof Let $m = n/d$. Then as in [80] we have

$$W_d(h) \leq \sum_{a=1}^{n/d-1} |T(ad, h)|^2 = mM - h^2,$$

where M is the number of solutions of the congruence

$$u \equiv v \pmod{m}, \qquad 1 \leq u, v \leq h.$$

If we express h in the form $h = km + r$ with $0 \leq r \leq m - 1$, then clearly, $M = r(k + 1)^2 + (m - r)k^2 = k^2m + 2kr + r$. Therefore

$$\begin{aligned} W_d(h) &\leq mM - h^2 = k^2m^2 + 2kmr + mr - h^2 \\ &= (h - r)^2 + 2r(h - r) + mr - h^2 = r(m - r). \end{aligned}$$

Taking into account that $r \leq h$, we obtain the desired statement.   □

Theorem 17.4 *The bound*

$$J(q, n, \Delta) \leq \frac{4n^3}{(\Delta - 1)^2 t}$$

holds.

Proof Let $h = \lfloor \Delta/2 \rfloor$ and let N_j denote the number of solutions of the congruence

$$jq^x \equiv l + h + u - v \pmod{n}, \qquad x = 1, \dots, t, \quad u, v = 1, \dots, h.$$

It is evident that $J(q, n, \Delta) \leq |\mathfrak{I}(q, n, \Delta)|$ where $\mathfrak{I}(q, n, \Delta)$ is the set of $j = 0, 1, \dots, n - 1$ for which this congruence is unsolvable, that is, $N_j = 0$. Set

$$S(a) = \sum_{x=1}^{t} e(aq^x/n).$$

Clearly, $N_j = th^2/n + R_j/n$ where

$$R_j = \sum_{a=1}^{n-1} S(aj)|T(a,h)|^2 \mathbf{e}(-a(l+h)/p).$$

Let us consider

$$R = \sum_{j=0}^{n-1} R_j^2.$$

We have

$$R = \sum_{j=0}^{n-1} \sum_{a,b=1}^{n-1} S(aj)S(bj)|T(a,h)|^2|T(b,h)|^2 \mathbf{e}\left(-(a+b)(l+h)/p\right)$$

$$= \sum_{a,b=1}^{n-1} |T(a,h)|^2|T(b,h)|^2 \mathbf{e}\left(-(a+b)(l+h)/p\right) \sum_{j=0}^{n} S(aj)S(bj).$$

Then,

$$\sum_{j=0}^{n-1} S(aj)S(bj) = \sum_{x,y=1}^{t} \sum_{j=0}^{n-1} \mathbf{e}\left(j(aq^x + bq^y)/n\right) = nN(a,-b).$$

For all divisors $d \mid n$ we gather together all terms corresponding to a and b with $\gcd(a,n) = \gcd(b,n) = d$. Applying Lemma 17.2, we obtain

$$R = n \sum_{\substack{d \mid n \\ d < n}} \sum_{\substack{\gcd(a,n)=d \\ \gcd(b,n)=d}} |T(a,h)|^2|T(b,h)|^2 N(a,-b)\mathbf{e}\left((a+b)(l+h)/p\right)$$

$$\leq nt \sum_{\substack{d \mid n \\ d < n}} dW_d(h)^2 \leq nt \max_{\substack{d \mid n \\ d < n}} dW_d(h) \sum_{\substack{d \mid n \\ d < n}} W_d(h).$$

From Lemma 17.3 and the identity

$$\sum_{\substack{d \mid n \\ d < n}} W_d(h) = \sum_{a=1}^{n-1} |T(a,h)|^2 = nh - h^2$$

we derive

$$R \leq n^3 h^2 t.$$

Since $R_j = -h^2 t$ for $j \in \mathfrak{I}(q,n,\Delta)$ then

$$|\mathfrak{I}(q,n,\Delta)|h^4 t^2 = \sum_{j=0}^{n-1} R_j^2 \leq n^3 h^2 t.$$

Taking into account that $h \geq (\Delta - 1)/2$, we obtain the result. $\qquad\square$

For certain values of parameters, the following statement provides a sharper bound.

Theorem 17.5 *The bound*

$$J(q, n, \Delta) \leq 2e^{1/2}n^{1-\alpha_q(\delta)}$$

holds, where $\delta = (\Delta - 1)/n$ *and*

$$\alpha_q(\delta) = \frac{\delta}{2\ln(3q/\delta)}.$$

Proof Let $\mathcal{J}$ be the half open interval

$$\mathcal{J} = \left[\frac{l+1}{n}, \frac{l+\Delta}{n}\right),$$

if $l + \Delta \leq n$, and let $\mathcal{J}$ be the union of the intervals

$$\left[\frac{l+1}{n}, 1\right) \quad \text{and} \quad \left[0, \frac{l+\Delta-n}{n}\right),$$

if $l + \Delta > n$. That is, in both cases $\mathcal{J}$ is a half open interval of length $(\Delta - 1)/n = \delta$ taken modulo 1.
Denote

$$r = \lfloor \log_q(3/\delta) \rfloor + 1,$$

thus $\delta q^r > 3$. Let us define the set

$$\mathcal{K} = \left\{k \; : \; \left[\frac{k}{q^r}, \frac{k+1}{q^r}\right) \subset \mathcal{J}\right\}.$$

The cardinality of this set $s = |\mathcal{K}|$ satisfies the inequality

$$s \geq \lfloor \delta q^r \rfloor - 1 \geq 0.5\delta q^r. \tag{17.2}$$

We see that if for some j the congruence (17.1) is unsolvable, then the q-ary expansion of j/n cannot contain the q-ary expansion of any $k \in \mathcal{K}$ as a substring (which is considered to be of the same length r, adding if necessary several zeros to the beginning).
Let us denote

$$m = \lfloor r^{-1} \log_q n \rfloor, \qquad T = mr.$$

Now we observe that the number of fractions j/n, $j = 1, \ldots, n$, such that the strings formed by the first T digits of their q-ary representations coincide, does not exceed $\lfloor n/q^T \rfloor + 1$.

To estimate $J(q, n, \Delta)$, we count the number N of strings of length T satisfying the following condition: for any $i = 0, \ldots, m - 1$, the substring formed by elements which stay on the positions

$$ir + 1, \ldots, (i + 1)r$$

differs from all s strings of length r, corresponding to elements of $\mathcal{K}$. Obviously, $N = (q^r - s)^m$. Therefore,

$$
\begin{aligned}
J(q, n, \Delta) \quad &\le \quad \left(\left\lfloor n/q^T \right\rfloor + 1 \right) N \le 2nq^{-T} (q^r - s)^m \\
&= \quad 2n \left(\frac{q^r - s}{q^r} \right)^m \le 2n \exp\left(-msq^{-r} \right).
\end{aligned}
$$

From (17.2) we see that

$$J(q, n, \Delta) \le 2n \exp\left(-0.5\delta m \right).$$

From the obvious inequality

$$r < \log_q (3/\delta) + 1 = \log_q (3q/\delta),$$

we see that

$$m \ge \frac{\log_q n}{r} - 1 \ge \frac{\log_q n}{\log_q (3q/\delta)} - 1 = \frac{\ln n}{\ln(3q/\delta)} - 1,$$

and the assertion follows. $\qquad\qquad\qquad\qquad\qquad\qquad\qquad\qquad\quad$ $\square$

Previously such bounds were known for primitive BCH codes only [5, 54].

Question 17.6. *Obtain upper bounds of $J(q, n, \Delta)$ in terms of the actual code distance rather than in terms of the designed distance.*

18

An Enumeration Problem in Finite Fields

For an integer $n > 1$ and a prime p, denote by $N_{n,p}(a)$ the number of solutions of the equation

$$\sum_{z=1}^{p-1} \vartheta_z z^n \equiv a \pmod{p},$$

in binary vectors $(\vartheta_1, \ldots, \vartheta_{p-1}) \in \{0, 1\}^{p-1}$.

It is proved in [71] that the estimate

$$N_{n,p}(a) = 2^{p-1} p^{-1} + \exp\left(O(np^{1/2} \ln p)\right)$$

holds.

Evidently the previous result is non-trivial for n of order at most $p^{1/2} \ln^{-1} p$. This bound has been improved in [83]. Using (3.20), we obtain the following results which improves [71, 83].

Theorem 18.1 *The bound*

$$|N_{n,p}(a) - 2^{p-1} p^{-1}| \le \begin{cases} \exp\left(O(np^{1/2} \ln p)\right); & \text{if } n \le p^{1/3}; \\ \exp(O(n^{5/8} p^{5/8} \ln p)); & \text{if } p^{1/3} \le n \le p^{1/2}; \\ \exp(O(n^{3/8} p^{3/4} \ln p)); & \text{if } p^{1/2} \le n \le p^{2/3}; \end{cases}$$

holds.

Proof Let us define

$$P(n, p) = \max_{\alpha=1,\ldots,p-1} \left| \prod_{z=1}^{p-1} \left(1 + \mathbf{e}(\alpha z^n / p)\right) \right|.$$

One sees that

$$
\begin{aligned}
N_{n,p}(a) &= \frac{1}{p} \sum_{\alpha=0}^{p-1} \sum_{(\vartheta_1,\ldots,\vartheta_{p-1})\in\{0,1\}^{p-1}} \mathbf{e}\left(\alpha\left(\sum_{z=1}^{p-1}\vartheta_z z^n - a\right)/p\right) \\
&= 2^{p-1}p^{-1} + \frac{1}{p} \sum_{\alpha=1}^{p-1} \sum_{(\vartheta_1,\ldots,\vartheta_{p-1})\in\{0,1\}^{p-1}} \mathbf{e}\left(\alpha\left(\sum_{z=1}^{p-1}\vartheta_z z^n - a\right)/p\right) \\
&= 2^{p-1}p^{-1} + \frac{1}{p} \sum_{\alpha=1}^{p-1} \mathbf{e}(-a\alpha/p) \prod_{z=1}^{p-1}\left(1 + \mathbf{e}(\alpha z^n/p)\right).
\end{aligned}
$$

Therefore,

$$
|N_{n,p}(a) - 2^{p-1}p^{-1}| \le P(n,p). \tag{18.1}
$$

Now we are going to estimate $P(n,p)$ in terms of the largest value of the Gaussian sums $G_n(p)$ which is defined by (3.19).

For N complex numbers $z_1, \ldots, z_N$ on the unit circle, $|z_1| = \cdots = |z_N| = 1$, we define

$$
P = \max_{|z|=1}\left|\prod_{k=1}^{N}(z+z_k)\right|, \qquad S = \max_{\nu=1,\ldots,N}\left|\sum_{k=1}^{N}z_k^\nu\right|.
$$

Let σ_i be the ith elementary symmetric function of $z_1, \ldots, z_N$; $\sigma_0 = 1$. Using Newton's formulas

$$
\sigma_i = \frac{1}{i}\sum_{j=1}^{i}(-1)^{j-1}\sigma_{i-j}\sum_{k=1}^{N}z_k^j, \qquad i = 1, \ldots, N,
$$

it is easy to prove by induction on i that

$$
|\sigma_i| \le \frac{S}{i}\prod_{m=1}^{i-1}(1+S/m); \qquad i = 1, \ldots, N.
$$

Hence,

$$
\begin{aligned}
P &\le \sum_{i=0}^{N}|\sigma_i| \le 1 + \sum_{i=1}^{N}\frac{S}{i}\prod_{m=1}^{i-1}(1+S/m) \\
&= \prod_{m=1}^{N}(1+S/m) = \exp\left(\sum_{m=1}^{N}\ln(1+S/m)\right).
\end{aligned}
$$

From the Stirling formula, we derive that

$$
\sum_{1\le m\le S}\ln(1+S/m) \le S\ln(2S) - \sum_{1\le m\le S}\ln m = O(S).
$$

Also, we have

$$\sum_{S<m\leq N} \ln(1 + S/m) = O\left(S\,\mathrm{Ln}(N/S)\right).$$

Therefore

$$P \leq \exp\left(O(S\,\mathrm{Ln}(N/S))\right). \tag{18.2}$$

Taking $N = p - 1$, $z_k = \mathbf{e}(k^n/p)$, $k = 1, \ldots, p - 1$, from (3.20) and (18.1) we obtain the desired result. $\qquad\square$

We remark that in the proof of Theorem 18.1 we use the bound (18.2) in a simplified form $P \leq \exp(S\,\mathrm{Ln}\,N)$. Applying this bound in the full strength one can obtain slightly sharper results.

For larger values of n, a weaker but still non-trivial result follows from the estimate (4.1). We also note that the method we used in the proof of Theorem 18.1 requires a bound of $G_n(p)$ which is at least $\ln p$ times better than the trivial one. In the following theorem we derive another inequality which produces a non-trivial estimate of $N_{n,p}(a)$ from very weak estimates of $G_n(p)$.

Theorem 18.2 *For any $\varepsilon > 0$ and*

$$p > n \ln n (\ln\ln n)^{-1+\varepsilon},$$

the bound

$$N_{n,p}(a) = 2^{p-1}p^{-1}\left(1 + O\left(\exp(-c(\varepsilon)p\ln^{-1+\varepsilon}p)\right)\right)$$

holds, where $c(\varepsilon) > 0$ depends on ε only.

Proof For any complex numbers $z, z_1, \ldots, z_N$ on the unit circle, $|z| = |z_1| = \cdots = |z_N| = 1$, we have

$$\left|\prod_{k=1}^{N}(z + z_k)\right|^2 = 2^N \prod_{k=1}^{N}[1 + \Re(zz_k)]$$
$$\leq 2^N\left(\frac{1}{N}\sum_{k=1}^{N}[1 + \Re(zz_k)]\right)^N$$
$$\leq 2^N\left(1 + \frac{1}{N}\left|\sum_{k=1}^{N}z_k\right|\right)^N.$$

From the bounds (4.1) and (18.1) and the previous inequality, we obtain the second statement of the theorem. $\qquad\square$

Finally, we remark that some interesting lower bounds for $P(n, p)$ have recently been found in [4].

Bibliography

[1] W. R. Alford, A. Granville and C. Pomerance, 'There are infinitely many Carmichael numbers', *Ann. Math.*, **140** (1994), 703–722.

[2] N. N. Andreev, S. V. Konyagin and A. Yu. Popov, 'Extremum problems for functions with small support', *Math. Notes*, **60** (1996), 241–247.

[3] R. C. Baker and G. Harman, 'Shifted primes without large prime factors', *Acta Arithm.*, **83** (1998), 331–361.

[4] J. P. Bell, P. B. Borwein and L. B. Richmond, 'Growth of the product $\prod_{j=1}^{n}(1 - x^{a_j})$', *Preprint*, 1998, 1–22.

[5] E. R. Berlekamp, *Algebraic Coding Theory*, McGraw-Hill, NY, 1968.

[6] L. Blum, M. Blum and M. Shub, 'A simple unpredictable pseudo-random number generator', *SIAM J. Comp.*, **15** (1986), 364–383.

[7] L. P. Bocharova, V. S. Van'kova and N. M. Dobrovolski, 'On the computation of optimal coefficients', *Matem. Zametki*, **49** (1991), 23–28 (in Russian).

[8] D. A. Clark and M. R. Murty, 'The Euclidean algorithm for Galois extensions of $\mathbb{Q}$', *J. Reine und Angew. Math.*, **459** (1995), 151–162.

[9] W. E. Clark and L. W. Lewis, 'Prime cyclic arithmetic codes and the distribution of power residues', *J. Number Theory*, **32** (1989), 220–225.

[10] J. H. Conway and N. J. A. Sloan, *Sphere Packings, Lattices and Groups*, Springer-Verlag, Berlin, 1993.

[11] H. Davenport, 'Linear forms associated with an algebraic number field', *Quart. Journ. Math.*, **3** (1952), 32–41.

[12] S. Egami, 'The distribution of residue classes modulo a in an algebraic number field', *Tsucuba J. of Math.*, **4** (1980), 9–13.

[13] J.-H. Evertse and H. P. Schlickewei, 'The absolute subspace theorem and linear equations with unknowns from a multiplicative group', *Preprint*, 1997, 1–22.

[14] J. Fabrykowski, 'On quadratic residues and nonresidues in difference sets modulo m', *Proc. Amer. Math. Soc.*, **36** (1994), 325–331.

[15] L. Flatto, J. C. Lagarias and A. D. Pollington, 'On the range of fractional parts of $\{\xi(p/q)^n\}$', *Acta Arithm.*, **70** (1995), 125–147.

[16] A. M. Frieze, J. Håstad, R. Kannan, J. C. Lagarias and A. Shamir, 'Reconstructing truncated integer variables satisfying linear congruence', *SIAM J. Comp.*, **17** (1988), 262–280.

[17] A. Garcia and J. F. Voloch, 'Fermat curves over finite fields', *J. Number Theory*, **30** (1988), 345–356.

157

[18] C. Garcia and P. Sole, 'Diameter lower bounds for Waring graphs and multiloop networks', *Discrete Math.*, **111** (1993), 257–261.

[19] J. von zur Gathen and I. Shparlinski, 'On orders of Gauss periods in finite fields', *Lect. Notes in Comp. Sci.*, **1004** (1995), 208–215.

[20] K. Girstmair, 'On the cosets of the $2q$-power group in the unit group modulo p', *Abh. Math. Sem. Univ. Hamburg*, **62** (1992), 217–232.

[21] K. Girstmair, 'The digits of $1/p$ in connection with class number factors', *Acta Arithm.*, **67** (1994), 381–386.

[22] K. Girstmair, 'Class number factors and residue class groups', *Abh. Math. Sem. Univ. Hamburg*, **67** (1997).

[23] D. M. Gordon, 'Equidistant arithmetic codes and character sums', *J. Number Theory*, **46** (1994), 323–333.

[24] S. Gupta and D. Zagier, 'On the coefficients of the minimal polynomials of Gaussian periods', *Math. Comp.*, **60** (1993), 385–398.

[25] S. Gurak, 'Factor of the period polynomials for finite fields, 1', *Contemp. Math.*, Amer. Math. Soc., **166** (1994), 309–333.

[26] S. Gurak, 'Factor of the period polynomials for finite fields, 2', *Contemp. Math.*, Amer. Math. Soc., **168** (1994), 127–138.

[27] S. Gurak, 'On the last factor of the period polynomials for finite fields', *Acta Arithm.*, **71** (1994), 391–400.

[28] S. Gurak, 'On the minimal polynomials for certain Gauss periods over finite fields', *Finite Fields and Applications, London. Math. Soc. Lect. Notes*, **233**, 1996, 85–96.

[29] K. Győry, 'On arithmetic graphs associated with integral domains', in *A Tribute To Paul Erdős*, Cambridge Univ. Press, 1990, 207–222.

[30] K. Győry, 'Some recent applications of "S-unit equations"', *Asterisque*, **209** (1992), 17–38.

[31] J. Håstad, J. C. Lagarias and A. M. Odlyzko, 'On the distribution of multivariate translates of sets of residues (mod p)', *J. Number Theory*, **209** (1994), 108–122.

[32] D. R. Heath-Brown, 'Artin's conjecture for primitive roots', *Quart. J. Math.*, **37** (1986), 27–38.

[33] D. R. Heath-Brown and S. Konyagin, 'New bounds for Gauss sums derived from kth powers, and for Heilbronn's exponential sum', *Quart. J. Math.* (to appear).

[34] J.-P. Kahane, *Séries de Fourier Absolument Convergentes*, Springer-Verlag, Berlin, 1970.

[35] A. A. Karatsuba, 'On bounds of complete exponential sums', *Matem. Zametki*, **1** (1967), 199–208 (in Russian).

[36] H. J. Karloff and P. Raghavan, 'Randomized algorithms and pseudorandom numbers', *J. ACM*, **40** (1993), 454–476.

[37] D. E. Knuth, *The Art of Computer Programming, vol. 2*, Addison-Wesley, Massachusetts, 1981.

[38] H. Kober, 'On the arithmetic and geometric means and on Hölder's inequality', *Proc. Amer. Math. Soc.*, **9** (1958), 452–459.

[39] H. Koblitz, *P-adic Numbers, P-adic Analysis, and Zeta-functions*, Springer-Verlag, Berlin, 1984.

[40] S. V. Konyagin, 'On the number of solutions of an univariate congruence of nth degree', *Matem. Sbornik*, **102** (1979), 171–187 (in Russian).

[41] S. V. Konyagin, 'On estimates of Gaussian sums and Waring problem modulo a prime', *Proc. Math. Inst. Russian Acad. Sci.*, Moscow, **198** (1992), 111–124 (in Russian).

[42] S. V. Konyagin and S.B. Stechkin , 'An estimate of the number of solutions of an univariate congruence of nth degree', *Proc. Math. Inst. Russian Acad. Sci.*, Moscow, **219** (1997), 249–257 (in Russian).

[43] N. M. Korobov, *Number–Theoretic Methods in Approximate Analysis*, Moscow, 1963 (in Russian).

[44] N. M. Korobov, 'On the distribution of digits in periodic fractions', *Matem. Sbornik*, **89** (1972), 654–670 (in Russian).

[45] N. M. Korobov, *Exponential Sums and Their Applications*, Kluwer Acad. Publ., North-Holland, 1992.

[46] J. C. Lagarias, 'Pseudorandom number generators in cryptography and number theory', *Cryptography and Number Theory, Proc. Symp. in Appl. Math.*, **42** (1990), 115–143.

[47] G. Larcher and H. Niederreiter, 'Optimal coefficients modulo prime powers in the three-dimensional case', *Annali di Mathematica pura ed applicata*, **155** (1989), 299–315.

[48] F. Lemmermeyer, 'The Euclidean algorithm in algebraic number fields', *Expos. Math.*, **13** (1995), 385–416.

[49] H. W. Lenstra, 'Euclidean number fields of large degree', *Invent. Math.*, **38** (1977), 237–254.

[50] H. W. Lenstra, 'On Artin's conjecture and Euclid's algorithm in global fields', *Invent. Math.*, **42** (1977), 201–224.

[51] A. J. Lichtman, 'The soluble subgroups and the Tits alternative in linear groups over rings of fractions of polycyclic group rings', *J. Pure Appl. Algebra*, **86** (1993), 231–287.

[52] R. Lidl and H. Niederreiter, *Finite Fields*, Cambridge University Press, Cambridge, 1994.

[53] P. Lisoněk, *Computer-Assisted Studies in Algebraic Combinatorics*, Ph.D. Thesis, Linz Univ., 1994, 1–217.

[54] F. J. MacWilliams and N. J. A. Sloane, *The Theory of Error-Correcting Codes*, North-Holland Publ. Comp., 1977.

[55] K. Mahler, 'An unsolved problem on powers of $3/2$', *J. Austral. Math. Soc.*, **8** (1969), 313–321.

[56] II.B. Mann, 'On linear relations between roots of unity', *Mathematika*, **12** (1965), 107–117.

[57] H. L. Montgomery, 'Distribution of small powers of a primitive root', in *Advances in Number Theory*, Clarendon Press, Oxford, 1993, 137–149.

[58] H. L. Montgomery, Personal Communication, 1993.

[59] H. L. Montgomery, R. C. Vaughan and T. D. Wooley, 'Some remarks on Gauss sums associated with kth powers', *Math. Proc. Cambr. Phil. Soc.*, **118** (1995), 21–33.

[60] M. R. Murty, 'Finitely generated groups (mod p)', *Proc. Amer. Math. Soc.*, **122** (1994), 37–45.

[61] M. R. Murty, M. Rosen and J. H. Silverman, 'Variations on a theme of Romanoff', *Intern. J. Math. Soc.*, **7** (1996), 373–391.

[62] G. Myerson, 'A combinatorial problem in finite fields, 1', *Pacif. J. Math.*, **82** (1979), 179–187.

[63] G. Myerson, 'A combinatorial problem in finite fields, 2', *Quart. J. Math.*, **31** (1980), 219–231.

[64] T. Nakahara, 'On a periodic solution of some congruence', *Rept. Fac. Sci. and Engin., Saga Univ., Math.*, **14** (1986), 1–5.

[65] W. Narkiewicz, *Elementary and Analytic Theory of Algebraic Numbers*, Polish Sci. Publ., Warszawa, 1990.

[66] W. Narkiewicz, *Polynomial Mappings*, Springer-Verlag, Berlin, vol. 1600, 1995.

[67] H. Niederreiter, 'Quasi-Monte Carlo methods and pseudo-random numbers', *Bull. Amer. Math. Soc.*, **84** (1978), 957–1041.

[68] H. Niederreiter, 'On a problem of Kodama concerning the Hasse–Witt matrix and distribution of residues', *Proc. Japan Acad.*, Ser.A, **63** (1987), 367–369.

[69] H. Niederreiter, *Random Number Generation and Quasi-Monte Carlo Methods*, SIAM Press, 1992.

[70] G. Niklash and R. Queme, 'An improvement of Lenstra's criterion for Euclidean number fields: the totally real case', *Acta Arithm.*, **58** (1991), 157–168.

[71] A. M. Odlyzko and R. P. Stanley, 'Enumeration of power sums modulo a prime', *J. Number Theory*, **10** (1978), 263–272.

[72] F. Pappalardi, 'On the order of finitely generated subgroups of $\mathbb{Q}^*$ (mod p) and divisors of $p - 1$', *J. Number Theory*, **57** (1996), 207–222.

[73] C. Powell, 'Bounds for multiplicative cosets over fields of prime order', *Math. Comp.*, **66** (1997), 807–822.

[74] K. Prachar, *Primzahlverteilung*, Springer-Verlag, Berlin, 1957.

[75] R. M. Robinson, 'Numbers having m small mth roots (mod p)', *Math. Comp.*, **61** (1993), 393–413.

[76] W. Schwarz and W. C. Waterhouse, 'The asymptotic density of supersingular Fermat varieties', *Arch. Math.*, **43** (1984), 142–144.

[77] A. Schinzel, 'Reducibility of Lacunary polynomials', *Acta Arithm.*, **50** (1988), 91–106.

[78] I. E. Shparlinski, 'Bounds for exponential sums with recurrent sequences and their applications', *Proc. Voronezh State Pedagogical Inst.*, **197** (1978), 74–85 (in Russian).

[79] I. E. Shparlinski, 'On residue classes modulo a prime number in an algebraic number field', *Matem. Zametki*, **43** (1988), 433–437 (in Russian).

[80] I. E. Shparlinski, 'On the dimension of BCH codes', *Problemy Peredachi Inform.*, **25**(1) (1988), 100–103 (in Russian).

[81] I. E. Shparlinski, 'On bounds of Gaussian sums', *Matem. Zametki*, **50** (1991), 122–130 (in Russian).

[82] I. E. Shparlinski, 'On Gaussian sums for finite fields and elliptic curves', *Proc. 1st French–Soviet Workshop on Algebraic Coding, Paris, 1991, Lect. Notes in Computer Sci.*, **537** (1992), 5–15.

[83] I. E. Shparlinski, 'On exponential sums with sparse polynomials and rational functions', *J. Number Theory*, **60** (1996), 233–244.

[84] I. E. Shparlinski, *Finite Fields: Theory and Computation*, Kluwer Acad. Publ., Dordrecht, The Netherlands, 1999.

[85] C. L. Siegel, 'The trace of totally positive and real algebraic integers', *Annals of Math.*, **46** (1945), 302–312.

[86] S. B. Stechkin, 'An estimate for Gaussian sums', *Matematicheskie Zametki*, **17** (1975), 342–349 (in Russian).

[87] O. Strauch, 'On distribution functions of $\{\xi(3/2)^n\}$ mod 1', *Acta Arithm.*, **81** (1997), 25–35.

[88] F. Thaine, 'On the p-part of the ideal class group of $\mathbb{Q}(\zeta_p + \zeta_p^{-1})$ and Vandiver's conjecture', *Michigan Math. J.*, **42** (1995), 311–344.

[89] F. Thaine, 'Properties that characterize Gaussian periods and cyclotomic numbers', *Proc. Amer. Math. Soc.*, **124** (1996), 35–45.

[90] M. Tompa, 'Lecture notes on probabilistic algorithms and pseudorandom
      generators', *Technical Report 91-07-05, Dept. of Comp. Sci. and Engin.*, Univ.
      of Washington, Seatle, 1991.
[91] I. M. Vinogradov, *Elements of Number Theory*, Dover Publ., New York, 1954.
[92] T. Washio and T. Kodama, 'Hasse–Witt matrices of hyperelliptic function fields',
      Sci. Bull. Fac. Education Nagasaki Univ., **37** (1986), 9–15.
[93] T. Washio and T. Kodama, 'A note on a supersingular function fields', *Sci. Bull.
      Fac. Education Nagasaki Univ.*, **37** (1986), 17–21.

Index

DATE DUE